The
Drawworks
and the Compound

Unit I, Lesson 6
First Edition

By Kate Van Dyke

Published by

PETROLEUM EXTENSION SERVICE
THE UNIVERSITY OF TEXAS AT AUSTIN
Division of Continuing & Innovative Education
Austin, Texas

Originally produced by

INTERNATIONAL ASSOCIATION
OF DRILLING CONTRACTORS
Houston, Texas

1995

Permissions

The following publishers have graciously granted permission to use photographs as they appear in other publications, to use photographs as the basis for new artwork, or to adapt already published material:

Diamond Chain Company
PO Box 561583
Dallas, Texas 75356
W. R. Evans-Lombe, southern area manager
John L. Wright, general product manager

Glencoe McGraw-Hill
PO Box 9609
Mission Hills, California 91346
Mark Schaefer, permissions coordinator

International Association of Oilwell Contractors
PO Box 4287
Houston, Texas 77210
Ken Fischer, director, Committee Operations

National Oilwell
1255 N. Post Oak Rd.
Houston, Texas 77055
Tom Harmon

Twin Disc Inc.
1328 Racine St.
Racine, Wisconsin 53403
David H. Johnson, manager of marketing
    communications

Library of Congress Cataloging-in-Publication Data

Van Dyke, Kate, 1951–
    The drawworks / by Kate Van Dyke. — 1st ed.
        p.   cm. — (Rotary drilling series ; unit I, lesson 6)
    "In cooperation with International Association of Drilling
Contractors, Houston, Texas."
    ISBN 0-88698-171-9
    1. Oil well drilling rigs. 2. Hoisting machinery.   I. University of Texas at
Austin. Petroleum Extension Service.   II. International Association of Drilling
Contractors.   III. Title.   IV. Series: Rotary drilling series; unit 1, lesson 6.
TN871.5.V27 1994
622'.3382—dc20   94-37140
                                                                              CIP

Catalog no. 2.106101
ISBN 0-88698-171-9

No state tax funds were used to publish this book. The University of Texas at Austin is an equal opportunity employer.

Contents

▼
▼
▼

Figures   v

Foreword   VII

Acknowledgments   IX

Units of Measurement   x

Introduction   I

The Hoisting System   2
    Derricks and Masts   4
    Blocks and Drilling Line   4
    Drawworks   5
    To summarize   6

Components of the Drawworks   7
    Frame   9
    Drum   9
    Catshaft   10
    Transmission   10
    Rotary Drive Countershaft   11
    Brakes   11
    To summarize   12

Getting Power to the Drawworks   13
    Mechanical and Electric Drives   13
    Comparison of Electric and Mechanical Drives   15
    To summarize   16

Transmissions   17
    Compounding Transmission   17
    Design of the Compound   21
    Selective Transmission   26
    To summarize   30
    Construction of Chains and Sprockets   31
    To summarize   40
    Installing Chain   41
    To summarize   46
    Lubrication of Chain-and-Sprocket Drives   47
    To summarize   52
    Maintenance of Chain-and-Sprocket Drives   53
    To summarize   60

Clutches   61
    Locations   61
    Positive Clutches   62
    Friction Clutches   64
    Overrunning Clutches   66
    Installation   67
    Maintenance   68
    To summarize   69

Main Brake   70
   Design   70
   Operation   73
   To summarize   78
   Maintenance   79
   Brake Flanges   82
   To summarize   84

Auxiliary Brake   85
   Hydrodynamic Brake   86
   Electrodynamic Brake   88
   To summarize   90

Catshaft   91
   Catheads   92
   Sand Reel   95
   To summarize   96

Lubrication   97
   To summarize   98

Glossary   99

Review Questions   111

Answers to Review Questions   123

Figures

▼
▼
▼

1. Windlass   1
2. Hoisting with a pulley and line   2
3. Hoisting system   3
4. Drawworks   5
5. Drawworks with guards removed   8
6. Drawworks drum   9
7. Prime movers and compound in a mechanical drive   14
8. Diesel engine and generator in an electric drive system   14
9. Compound for two engines   15
10. Hydraulic torque converter   18
11. Waterwheel   19
12. Hydraulic coupling and torque converter   20
13. View of the compound from above   21
14. Sprocket   22
15. Chain-and-sprocket drive   22
16. Multistrand drives   24
17. Gears 26
18. Power flow diagram of the selective transmission   28, 29
19. Roller links and pin links   31
20. Roller chain   32
21. Connector links   33
22. Offset link   33
23. Multistrand chain   34
24. Dimensions of a link   35
25. Broken pin   37
26. Broken link plate   37
27. Chain case and chain guard   39
28. Measuring the catenary on a chain   42
29. Chain with too much slack   43
30. Angular misalignment   44
31. Offset misalignment   45
32. Checking for misalignment and levelness of shafts   46
33. Cross section of two-strand chain showing lubrication   47
34. Drip lubrication   48
35. Oil bath lubrication   49
36. Slinger disk lubrication   49
37. Pressure lubrication   50
38. Best location to apply lubrication   50
39. Measuring chain for elongation   55
40. Worn sprockets   55
41. Positive clutches   62
42. Friction clutches   65
43. Overrunning clutch   66
44. Main (mechanical) brake   71

45. Braking capacity increases from the live end to the dead end   73

46. Comparing the diameter of the drum to the rims   74

47. Comparing the width of the brake band   75

48. Angle of wrap of 270°   75

49. Cross section of the brake rim   76

50. Checking the band for roundness   80

51. Measuring scoring in the brake rim   82

52. Hydrodynamic brake   86

53. Electrodynamic brake   88

54. The catshaft, catheads, and sand (coring) reel from the rear of the drawworks   91

55. Air tugger   93

56. Grease fitting locations   97

Tables

1. Common chain sizes   36

2. Maximum chain speeds for different lubrication methods   48

3. Elongation limits for chain   54

4. Roller chain troubleshooting guide   57

Foreword

For many years, the Rotary Drilling Series has oriented new personnel and further assisted experienced hands in the rotary drilling industry. As the industry changes, so must the manuals in this series reflect those changes.

The revisions to both text and illustrations are extensive. In addition, the layout has been "modernized" to make the information easy to get; the study questions have been rewritten; and each major section has been summarized to provide a handy comprehension check for the student.

PETEX wishes to thank industry reviewers—and our readers—for invaluable assistance in the revision of the Rotary Drilling Series. On the PETEX staff, Deborah Caples designed the layout; Doris Dickey proofread innumerable versions; Sheryl Horton saw production through from idea to book; Ron Baker served as content editor for the entire series.

Although every effort was made to ensure accuracy, this manual is intended to be only a training aid; thus, nothing in it should be construed as approval or disapproval of any specific product or practice.

Kathy Bork

Acknowledgments

The revising of the 1982 edition of *The Hoist* into this new edition entitled *The Drawworks and the Compound* was a challenging journey. My thanks to several people who generously gave of their time and expertise and provided drawings, photos, and written resources: Ken Fischer of IADC; Wes Morrow of National Oilwell; Otis Danielson, retired; Dick Evans-Lombe of Diamond Chain Co.; and Kent Greenwald of Twin Disc.

Thanks also to eagle-eyed reviewers of the draft: John Altermann of Reading and Bates, Jim Arnold of Salem Investment, Joey Hopewell of Delta Drilling, and Ken Fischer.

Finally, thanks to the director of PETEX, Ron Baker for his good-humored explanations of convoluted reference material, to Sheryl Horton for providing referrals and acting as a sounding board, to Jonell Clardy for new illustrations, and to Terry Gregston for new photographs for this edition.

Kate Van Dyke

Units of Measurement

Throughout the world, two systems of measurement dominate: the English system and the metric system. Today, the United States is almost the only country that employs the English system.

The English system uses the pound as the unit of weight, the foot as the unit of length, and the gallon as the unit of capacity. In the English system, for example, 1 foot equals 12 inches, 1 yard equals 36 inches, and 1 mile equals 5,280 feet or 1,760 yards.

The metric system uses the gram as the unit of weight, the metre as the unit of length, and the litre as the unit of capacity. In the metric system, for example, 1 metre equals 10 decimetres, 100 centimetres, or 1,000 millimetres. A kilometre equals 1,000 metres. The metric system, unlike the English system, uses a base of 10; thus, it is easy to convert from one unit to another. To convert from one unit to another in the English system, you must memorize or look up the values.

In the late 1970s, the Eleventh General Conference on Weights and Measures described and adopted the Système International (SI) d'Unités. Conference participants based the SI system on the metric system and designed it as an international standard of measurement.

The *Rotary Drilling Series* gives both English and SI units. And because the SI system employs the British spelling of many of the terms, the book follows those spelling rules as well. The unit of length, for example, is *metre*, not *meter*. (Note, however, that the unit of weight is *gram*, not *gramme*.)

To aid U.S. readers in making and understanding the conver-

English-Units-to-SI-Units Conversion Factors

Quantity or Property	English Units	Multiply English Units By	To Obtain These SI Units
Length, depth, or height	inches (in.)	25.4	millimetres (mm)
		2.54	centimetres (cm)
	feet (ft)	0.3048	metres (m)
	yards (yd)	0.9144	metres (m)
	miles (mi)	1609.344	metres (m)
		1.61	kilometres (km)
Hole and pipe diameters, bit size	inches (in.)	25.4	millimetres (mm)
Drilling rate	feet per hour (ft/h)	0.3048	metres per hour (m/h)
Weight on bit	pounds (lb)	0.445	decanewtons (dN)
Nozzle size	32nds of an inch	0.8	millimetres (mm)
Volume	barrels (bbl)	0.159	cubic metres (m³)
		159	litres (L)
	gallons per stroke (gal/stroke)	0.00379	cubic metres per stroke (m³/stroke)
	ounces (oz)	29.57	millilitres (mL)
	cubic inches (in.³)	16.387	cubic centimetres (cm³)
	cubic feet (ft³)	28.3169	litres (L)
		0.0283	cubic metres (m³)
	quarts (qt)	0.9464	litres (L)
	gallons (gal)	3.7854	litres (L)
	gallons (gal)	0.00379	cubic metres (m³)
	pounds per barrel (lb/bbl)	2.895	kilograms per cubic metre (kg/m³)
	barrels per ton (bbl/tn)	0.175	cubic metres per tonne (m³/t)
Pump output and flow rate	gallons per minute (gpm)	0.00379	cubic metres per minute (m³/min)
	gallons per hour (gph)	0.00379	cubic metres per hour (m³/h)
	barrels per stroke (bbl/stroke)	0.159	cubic metres per stroke (m³/stroke)
	barrels per minute (bbl/min)	0.159	cubic metres per minute (m³/min)
Pressure	pounds per square inch (psi)	6.895	kilopascals (kPa)
		0.006895	megapascals (MPa)
Temperature	degrees Fahrenheit (°F)	$\dfrac{°F - 32}{1.8}$	degrees Celsius (°C)
Thermal gradient	1°F per 60 feet	—	1°C per 33 metres
Mass (weight)	ounces (oz)	28.35	grams (g)
	pounds (lb)	453.59	grams (g)
		0.4536	kilograms (kg)
	tons (tn)	0.9072	tonnes (t)
	pounds per foot (lb/ft)	1.488	kilograms per metre (kg/m)
Mud weight	pounds per gallon (ppg)	119.82	kilograms per cubic metre (kg/m³)
	pounds per cubic foot (lb/ft³)	16.0	kilograms per cubic metre (kg/m³)
Pressure gradient	pounds per square inch per foot (psi/ft)	22.621	kilopascals per metre (kPa/m)
Funnel viscosity	seconds per quart (s/qt)	1.057	seconds per litre (s/L)
Yield point	pounds per 100 square feet (lb/100 ft²)	0.48	pascals (Pa)
Gel strength	pounds per 100 square feet (lb/100 ft²)	0.48	pascals (Pa)
Filter cake thickness	32nds of an inch	0.8	millimetres (mm)
Power	horsepower (hp)	0.75	kilowatts (kW)
Area	square inches (in.²)	6.45	square centimetres (cm²)
	square feet (ft²)	0.0929	square metres (m²)
	square yards (yd²)	0.8361	square metres (m²)
	square miles (mi²)	2.59	square kilometres (km²)
	acre (ac)	0.40	hectare (ha)
Drilling line wear	ton-miles (tn•mi)	14.317	megajoules (MJ)
		1.459	tonne-kilometres (t•km)
Torque	foot-pounds (ft•lb)	1.3558	newton metres (N•m)

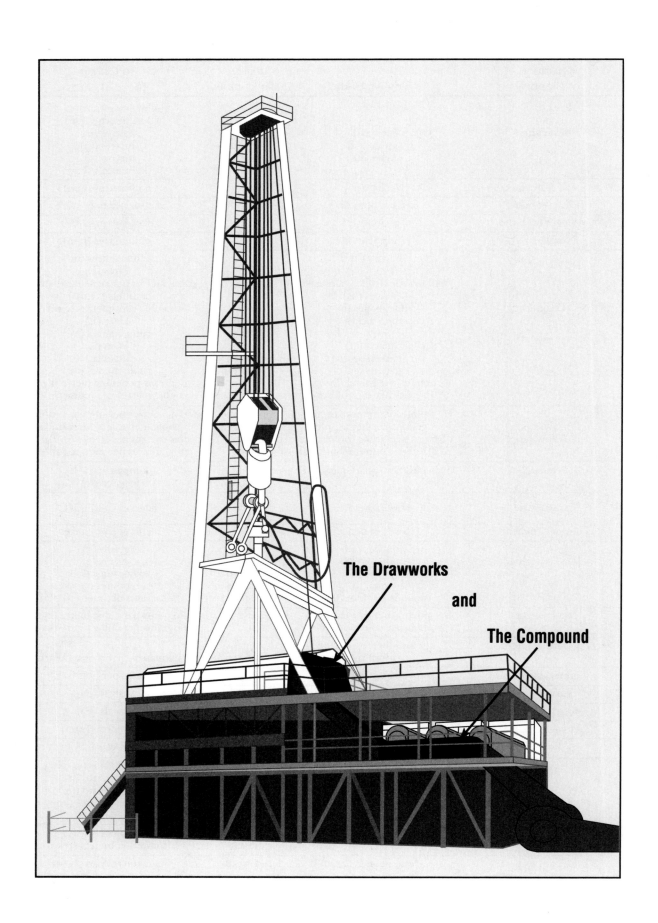

The Drawworks

and

The Compound

Introduction

▼
▼
▼

The drawworks is a part of the system that rotary drilling rigs use for hoisting, or lifting, the drill stem and casing out of the hole. The earliest hoist, a windlass or winch, was a simple drum, or spool, sitting horizontally between two posts with one end of a rope attached to it (fig. 1). The other end of the rope was attached to something a person wanted to lift, such as a bucket. When someone turned the drum with a crank, the rope wound around the drum and lifted the bucket. The windlass enabled people to lift a heavy load of water, for instance, much more easily than they could have by pulling the bucket straight up. Of course, enterprising laborers were always looking for power greater than human strength to turn the drum and hoist heavier loads, and they used animals and, eventually, engines for the purpose.

Early rigs used steam engines to power the hoist. Today they use diesel engines and electric motors. But the basic principle of using a mechanical device to do the work of lifting continues to be the basis of hoisting.

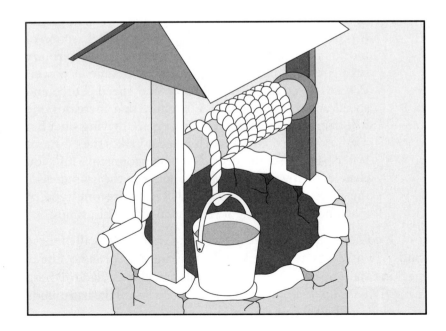

Figure 1. Windlass

The Hoisting System

The hoisting system of a drilling rig is a collection of machines that work together. Broadly speaking, a simple machine does one of the following things:

1. It converts energy from one form to another.

   Energy comes in several forms. Everything that moves has mechanical energy, and mechanical energy is the type most important to the hoisting system. Some other types of energy are heat, light, and electricity. An example of a machine that converts energy is a generator, which turns mechanical energy into electricity.

2. It transfers energy from one place to another.

   The steering wheel and its linkages are a machine that transfers mechanical energy (motion) from the driver's arm to the wheels of a car.

3. It controls energy.

   A machine can control mechanical energy in three ways to make it more usable: (a) switch it on and off. A clutch in a car interrupts the power from the engine to the wheels so that you don't have to turn off the engine at a stoplight; (b) change its direction. A pulley changes the direction of motion of a rope from linear (in a straight line) to rotary (in a circle) and back again (fig. 2); (c) change its power. Power is a combination of force and speed. Force and speed are always related—a machine that increases one will decrease the other. When you are driving on a flat road, stepping on the gas in high gear makes the car move faster—increases its speed. But when going up a hill, you change to a lower gear so that stepping on the gas increases the force instead of the speed. Gears are one type of machine for changing force and speed relationships.

A complex machine, like the hoisting system, does all of these kinds of work. It helps to think of the hoisting system as a complex machine made up of several other complex machines. Each of these in turn is made up of simpler machines that do one of the basic kinds of work described above.

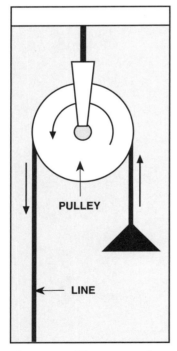

Figure 2. Hoisting with a pulley and line

Figure 3 shows the main components of the hoisting system. The derrick or mast supports the crown block and the traveling block, which are each a set of pulleys. The derrick or mast also supports the drill stem by means of the drilling hook hanging from the bottom of the traveling block. The drawworks, also called the hoist, sits on the ground or the rig floor underneath the derrick. It contains a drum corresponding to the drum of the windlass. A rope woven from wire and called the drilling line connects the drawworks drum to the blocks and does the same job as the rope on a windlass.

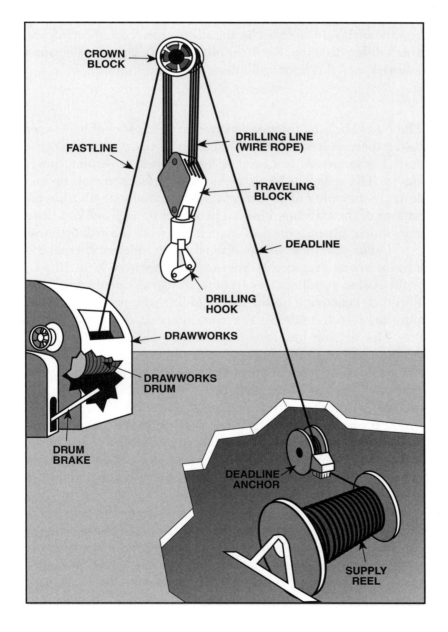

Figure 3. Hoisting system (derrick not shown)

Derricks and Masts

The derrick or mast, a steel tower that may rise 142 feet (43 metres) above the rig floor, is the universally recognized symbol of oilwell drilling. Derricks and masts look much the same and do the same job, but they are used at different types of drilling sites. A derrick is a more or less permanent structure. Its legs sit on the four corners of the rig floor, and the crew must disassemble it to move it. A mast, on the other hand, is portable. It fits into an A-frame that may sit on the rig floor or on the ground. The crew can fold or telescope it down to move it. Both a derrick and a mast must support many tons of weight of the drill stem, so they are very strong. The derrick—except as found on most offshore drilling rigs—has become fairly rare, the mast having all but replaced it. A derrick and a mast differ, of course, but in the oil industry, everyone calls a mast a derrick, and this book will follow that convention.

Blocks and Drilling Line

The *crown block* and the *traveling block* are each a set of pulleys, or *sheaves* (pronounced "shivs"). The crown block sits at the top of the derrick (the crown) and, except for the sheaves turning, never moves. The traveling block, as its name implies, travels up and down, in the center of the derrick. A *drilling hook* extends from the bottom of the traveling block. The swivel or a special top-drive unit, which suspends the drill stem, hangs from the drilling hook.

Drilling line looks like an ordinary fiber rope, except that it is made of woven steel wire. It ranges in diameter from ⅞ to 2 inches, or about 22 to 51 millimetres. It comes on a spool called a *supply reel*. This reel, depending on the length of line wrapped on it, can be quite large—6 feet (almost 2 metres) in diameter.

The drilling line runs from the supply reel through the deadline anchor to the crown block and passes over one sheave. Then it goes down to the traveling block and wraps around one of its sheaves and heads back up to the crown block. To multiply the strength of the hoisting system, and therefore the amount of weight it can hoist, the crew threads, or *reeves*, the line back and forth several times between the two blocks. Finally the end of the line coming from the crown block goes to the drawworks drum, where it is anchored. The drum spools the line in or out, thus lifting or lowering the traveling block.

The part of the drilling line running from the drawworks drum to the crown block is called the *fastline* because it moves rapidly on and off the drum. The part of the drilling line from the crown block to the deadline anchor is the *deadline*. It does not move at all during hoisting. The deadline anchor is fastened to the rig's substructure and secures the deadline with a clamping mechanism.

The drawworks, or hoist, houses the drum that the drilling line wraps around (fig. 4). It is one of the largest and heaviest pieces of equipment on a drilling rig. It also contains machinery to control the drum—brakes, clutches, and a transmission. The driller controls the whole hoisting system from a console on the drawworks. The drawworks and its power system are the most complex machines in the hoisting system.

Drawworks

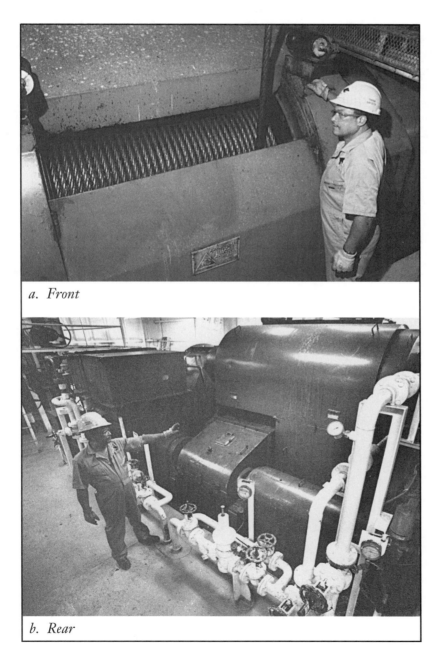

a. Front

b. Rear

Figure 4. Drawworks

To summarize—

Functions of a machine
- Converts energy from one form to another
- Transfers energy from one place to another
- Controls the direction, force, or speed of mechanical energy or turns it on and off

Components of the hoisting system
- Derrick or mast—a tower that supports the blocks and drill stem
- Drilling line—a wire rope that runs from a supply spool through the blocks to the drum
- Blocks—pulleys that the drilling line threads through; the crown block is stationary and the traveling block moves up and down
- Drawworks—a machine that houses and controls the drum which spools drilling line in and out

Components of the Drawworks

▼
▼
▼

ost of the rest of this book will describe the components of the drawworks, how they are constructed and powered, and what they do. The people who work with the drawworks must be able to operate and maintain it well so that the machinery lasts as long as possible and is safe to use.

Figure 5a shows a front view of a drawworks, with the coverings of the frame removed:

(1) driller's console and brake lever
(2) low drum drive
(3) main brake
(4) drum
(5) high drum drive
(6) catshaft and optional sand reel
(7) rotary drive countershaft (optional)
(8) auxiliary brake

The back of the drawworks is sometimes called the power side because it is the side nearest the engines that supply power to the drawworks (see Getting Power to the Drawworks). Figure 5b shows the uncovered back of the drawworks:

(1) high drum drive
(2) electric motors
(3) drum
(4) optional sand reel
(5) catshaft drive
(6) low drum drive
(7) output shaft
(8) input shaft

The chains in the rear are part of the transmission.

a. Front view

(1) driller's console and brake lever
(2) low drum drive
(3) main brake
(4) drum
(5) high drum drive
(6) catshaft and optional sand reel
(7) rotary drive countershaft (optional)
(8) auxiliary brake

1) high drum drive
(2) electric motors
(3) drum
(4) optional sand reel
(5) catshaft drive
(6) low drum drive
(7) output shaft
(8) input shaft

b. Rear view

Figure 5. Drawworks with guards removed

Frame

The frame of a drawworks is a strong and rigid platform constructed of heavy steel beams. A metal housing in the form of several guards fits over and encloses the components of the drawworks. The crew can remove the guards to inspect or service it. During operation, the guards protect workers and confine an oil spray that lubricates the transmission.

Manufacturers design drawworks to be movable from one drill site to another, so the frame is often small enough to transport on the highway. Sometimes, however, the drawworks is too large to move in one piece. In this case, it can split into two halves—main drum on the front half, transmission and catshaft on the back.

Drum

The heart of the drawworks is the drum (fig. 6). The drum is a huge steel spool that rotates on an axle called the drumshaft. When the driller feeds power to the drum, it spools up drilling line, which raises the traveling block, thereby hoisting the drill stem and casing. The drum brake rims can be about 3 to 5 feet (1 to 1½ metres) in diameter; the drum spool may vary from 1½ to 3 feet (½ to 1 metre). The deeper the well is, the longer the drilling line is, and the larger the diameter of the drum needs to be to spool it all up.

The rims on each end of the drum are wide and strong because the main brake grips them to stop the drum.

The outside surface of the spool has grooves in it. These grooves guide the drilling line as it winds around the drum so that it wraps evenly and continuously. This keeps the drum balanced and prevents the fastline from whipping back and forth in the air between the drum and the crown block when spooling at high speeds.

The manufacturer flame-hardens the rims and spool of the drum to resist wear. This process hardens an outer layer of the steel.

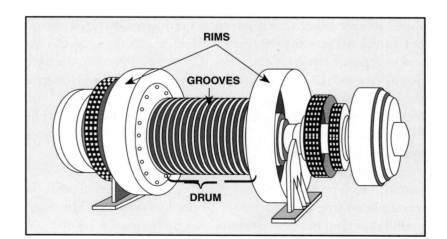

Figure 6. Drawworks drum

9

Most of the other components of the drawworks are there to control the drum: the brakes stop it, and a transmission and clutch system transfers the power to make it hoist. The driller controls the brakes and the transmission from a panel, or console, on the left side (see fig. 5a) of the drawworks (looking at it from the front). Rig workers refer to this side as the driller's side and the other side as the off-driller's side.

Catshaft

The drawworks has a shaft across the top called a *catshaft* that sticks out of each end (see fig. 5b). The catshaft may have a smaller drum called a *sand reel* on it, but its main purpose is to be an axle for small reels called *catheads*. A drawworks has four catheads. Two help the crew make up and break out the drill pipe, and two help to move heavy pieces of equipment around the rig floor. The sand reel is a light-duty hoist to raise and lower tools into and out of the hole.

Transmission

"Transmission" is just a general name for a machine that transmits mechanical energy, or motion, from an engine to a part that needs it. For example, the transmission in a car transmits power from the engine to one of the axles, which turns the wheels. Usually when we speak of a transmission, we are referring to a complex machine made up of many smaller machines that each transmit power from one particular spinning shaft to another. These smaller machines are called *drives*.

A transmission inside the drawworks directs power where it needs to be: to the drum, to the catshaft, or to the rotary table countershaft. When the power goes to the drum, the transmission splits into two drives, which also allow the driller to vary the *amount* of power, and therefore the speed at which the traveling block rises and how much weight it can hoist. These two drives are the high drum drive and the low drum drive. They are like high and low gear in a car. The low drum drive works best when hoisting a heavy load, like using low gear to go up a steep hill; it is on the driller's side. The high drum drive transmits less power and a faster speed; it is opposite the low drum drive.

Some rigs also have another type of transmission, the compound, which transfers power from the engines to the drawworks transmission (see Getting Power to the Drawworks). The compound is behind the drawworks on the rig floor.

Most drawworks include a rotary drive countershaft, which is a rotating shaft that gets power from one of the transmissions and sends it to the rotary table (see fig. 5a). It sits inside its own housing, which may be a part of the main drawworks frame or may be detachable for transportation.

The rotary drive countershaft is on the opposite side of the drawworks from the driller's console. For this reason, another name for the off-driller's side is the rotary side.

Rotary Drive Countershaft

The *main brake* consists of two bands fitted with brake pads; the bands fit over the two rims of the drum (see fig. 5b). When the driller engages this brake, the pads press down on the rims to stop the drum from hoisting or from letting out drilling line. The main brake also keeps the drum from rotating (and therefore holds the drill stem stationary) when making up or breaking out drill pipe.

When the driller releases the main brake, gravity pulls the traveling block and drill stem down and causes the drilling line to spool out and the drum to rotate. The drill stem may weigh more than 100 tons (91 tonnes), and it can drop very fast and be very hard to stop. So the drawworks drum also has an *auxiliary brake* (see fig. 5a). The auxiliary brake is always working when the block is descending, to prevent it from falling full-speed. It helps the main brake to slow and stop the block. The auxiliary brake may be either hydrodynamic (activated by water) or electrodynamic (activated by electricity).

The sand reel has its own brake that works like the main brake on the drum.

Brakes

To summarize—

Components of the drawworks

- Frame
- Drum
- Catshaft
- Transmission
- Rotary drive countershaft
- Brakes
- Driller's console

Job of each component

- Frame—encloses and supports the other components
- Drum—spools up drilling line and raises the traveling block and drill stem
- Catshaft—an axle for the catheads and sand reel
- Transmission—transmits power to the drum, catshaft, or rotary table countershaft
- Rotary drive countershaft—transmits power to the rotary table
- Brakes—stops the drum; the drawworks has a main brake and an auxiliary brake
- Driller's console—a panel with controls for the brakes and transmission

Getting Power to the Drawworks

Power for running the drawworks, and therefore the whole hoisting system, comes from the rig's *prime movers*. The prime movers are the basic power source. Most rigs use two to four internal-combustion engines as prime movers. These are the same type of engine that a car has, but rig engines are much bigger and more powerful. Most use diesel fuel because a diesel engine has more turning power, or *torque*, than a gasoline engine. Torque is important because the engine directly spins a shaft to produce mechanical energy that powers everything else.

Engine power is then transmitted to run the various working parts of the rig in one of two ways. A *mechanical-drive* rig uses a compound, a mechanical transmission made up of sprockets and chains (fig. 7). On an *electric-drive* rig, the prime movers drive generators mounted right on each diesel engine (fig. 8). A generator converts the mechanical energy from the engine shaft into electricity. The electricity then flows through electric cables to electric motors attached to the drawworks and other parts of the rig that need power. Each motor directly powers one of these parts (see electric motors in fig. 5b).

Mechanical and Electric Drives

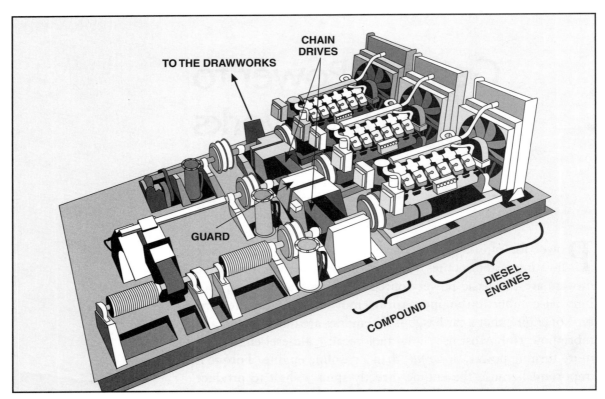

Figure 7. Prime movers and compound in a mechanical drive

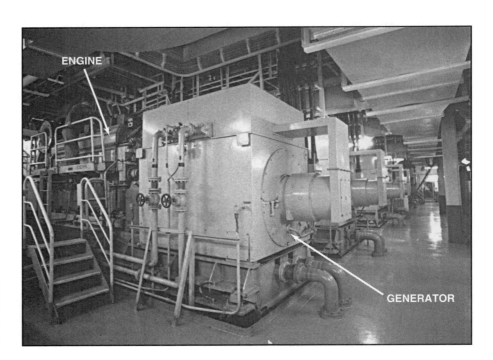

Figure 8. Diesel engine and generator in an electric drive system

About two thirds of rigs in operation today are mechanical. An electric drive has many advantages over a mechanical drive, however. First, it is much simpler. The more complex a machine is, the more things can go wrong with it and the more maintenance it needs. The mechanical transmission that a mechanical-drive rig needs is a huge, complex machine made up of many chain-and-sprocket drives (see Compound and fig. 9). The chain-and-sprocket drives are very large and heavy. A single drive may be 8 feet (about 2½ metres) long with chain links that are nearly as big as your fist. In order for all these drives to work properly and safely, the crew must lubricate, inspect, and periodically repair or replace all their metal parts, which is time-consuming and expensive.

A mechanical drive needs such a complex transmission because the driller may have to combine the power of several engines to lift heavy loads. Also, while two engines, for example, are powering the drawworks, a third may be powering the mud pumps.

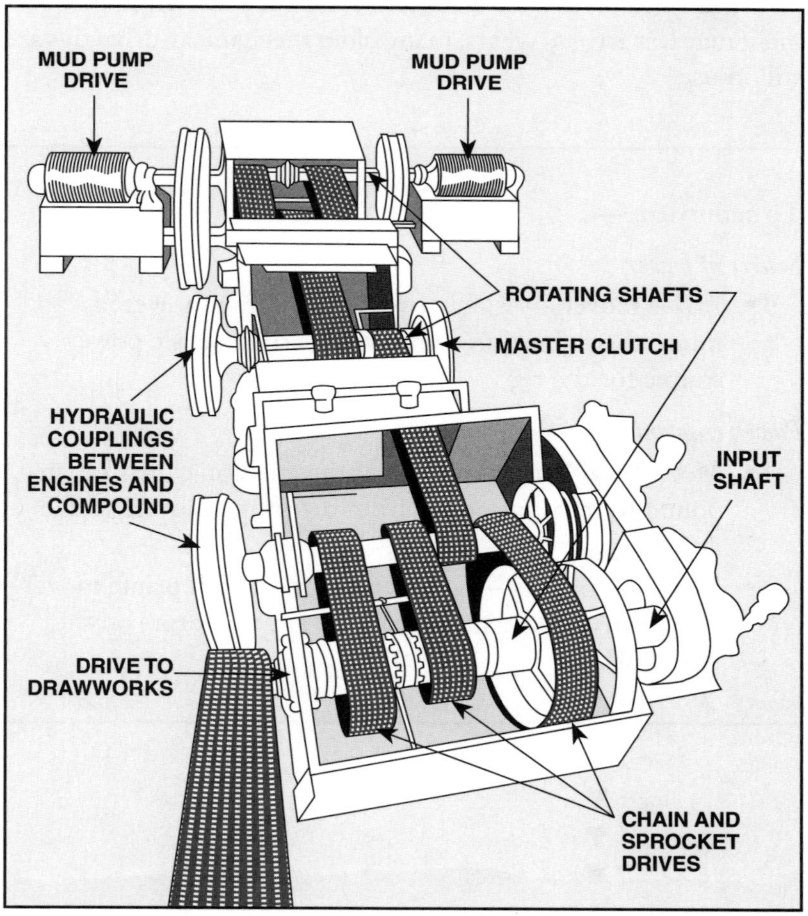

Comparison of Electric and Mechanical Drives

Figure 9. *Compound for two engines*

An electric drive eliminates all the chain-and-sprocket drives and replaces them with electric cables, which are much smaller and lighter and therefore easier to handle. Unlike the mechanical drive, an electric drive needs minimal maintenance. Also, with an electric drive, the driller can vary the speed and the force of electric motors continuously with a switch, in the same way you can vary the power to a lamp with a dimmer switch.

The second advantage of an electric drive is that its electric cables can be very long. A chain-and-sprocket drive cannot span an unlimited distance, so the prime movers must be near the machinery they are driving on the rig floor. The vibration may stress the equipment: for example, parts can get out of alignment and wear out faster. The cables that connect the generators of an electric drive to the motors next to the driven components can span a much longer distance than chains, so the engines of electric-drive rigs can be much farther from the rig floor, thereby lessening noise and eliminating vibration.

Many new rigs have electric drives, but since drilling equipment may last 20 or 30 years, many older mechanical-drive rigs are still in use.

To summarize—

Source of power
- Prime movers—usually two to four diesel-powered internal-combustion engines that are the basic power source for the rig

Power transmission
- Mechanical-drive rig—uses a transmission called a compound to transmit power from the prime movers to the drawworks
- Electric-drive rig—has a generator on each prime mover and electric cables to transmit power to motors on the drawworks

---

Transcriptions

Rotating shafts are the basis of a transmission. The rest of the transmission consists of an arrangement of mechanical parts, such as gears, sprockets and chains, belts and pulleys, and clutches, that connect the rotating shafts to each other. All together these parts constitute a machine that transmits power from a power source to another machine to make it work.

On a mechanical-drive rig, a *compounding transmission*, or *compound*, sends power from the engines to the drawworks and the rotary table, and sometimes to the mud pumps. Electric-drive rigs, where the engines run generators, do not have a compound. Here, the cables transmit electric power to motors near the components that need it.

Once the drawworks receives power from either the compound or electric motors, it must have its own system for sending that power to its various parts. A *selective transmission* does this job, allowing the driller to select how the power is distributed (torque/speed combinations) to various components of the drawworks. The drawworks on both mechanical-drive and electric-drive rigs have a selective transmission.

Many rigs need more than one prime mover to provide enough power, so in a mechanical-drive rig, the compounding transmission combines, or compounds, the energy from two to four engine shafts to make them act as one power source (fig. 9). The output shafts from the engines do not connect directly to the compound, however. Between each rotating shaft and the compound is either a torque converter or a hydraulic coupling to smooth the transfer of power.

Compounding Transmission

*Torque Converters and
Hydraulic Couplings*

In a car with a standard transmission, if the driver lets the clutch out too fast when starting in first gear, the car will jump forward and the engine may die. In order to smoothly start moving, the driver must let the clutch out slowly. Rig engines require the same type of transition of power, but they are too big for a person to control with a clutch. To get the power from the engines to the compound smoothly, each engine has a *torque converter* or a *hydraulic coupling* (fig. 10) instead of a clutch.

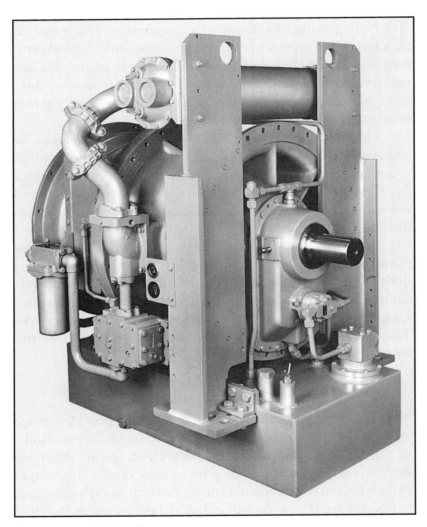

Figure 10. Hydraulic torque converter

A torque converter is a type of hydraulic coupling. The term "hydraulic" just means that it uses a liquid to work. A waterwheel is a simple type of hydraulic coupling. The liquid (water) pushes against vanes inside the rim of the wheel, which causes the wheel to rotate (fig. 11). A hydraulic coupling in a transmission system uses the principle of liquid pushing against a wheel in a different way. Whereas a waterwheel is using water to *supply* power to turn the wheel, a hydraulic coupling is using water to *resist* the rotation of a wheel that the engine shaft is already driving.

To understand how water resists the rotation of a wheel, think of pushing a canoe paddle through water. If you swing the paddle from the air down into the water with all your strength and speed, it will move as fast as you can swing your arm until it hits the water. Then the water resists your effort and the paddle moves smoothly and more slowly through the water. In fact, it is impossible to jerk the paddle through water the way you can in air; water always resists your effort and smooths the stroke. The liquid in a hydraulic coupling is even more viscous, or thicker and slower moving, than water. (Oil is an example of a liquid that is more viscous than water.) So it provides even more resistance to movement than water.

A hydraulic coupling has two halves that fit together and are filled with liquid (fig. 12a). One half consists of a hemispherical housing with an impeller inside of it. The impeller has vanes that are attached to the inside circumference of the housing. This side of the hydraulic coupling fits onto the engine's output shaft and turns at the same speed as the engine. When the engine is on, the vanes of the impeller rotate and push against the liquid and start it rotating.

The other half of the hydraulic coupling consists of a wheel with vanes called a turbine and its hemispherical housing. The turbine turns freely inside its housing and fits onto an output shaft that goes to the compound. The rotating liquid turns the turbine.

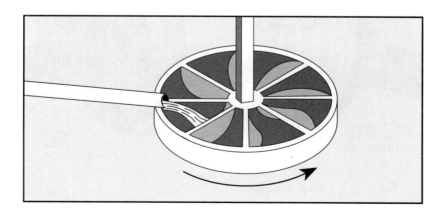

Figure 11. Waterwheel

In this way, the power is transmitted from the input shaft to the output shaft, but the liquid also absorbs the shock by resisting the movement.

A torque converter works in the same way as an ordinary hydraulic coupling except that it multiplies the torque, or turning power, of the engine at the same time. A third vaned wheel called a stator is what accomplishes this (fig. 12b). The stator causes the liquid returning from the turbine to push the impeller in the same direction it is already going, thus helping it to turn and increasing torque.

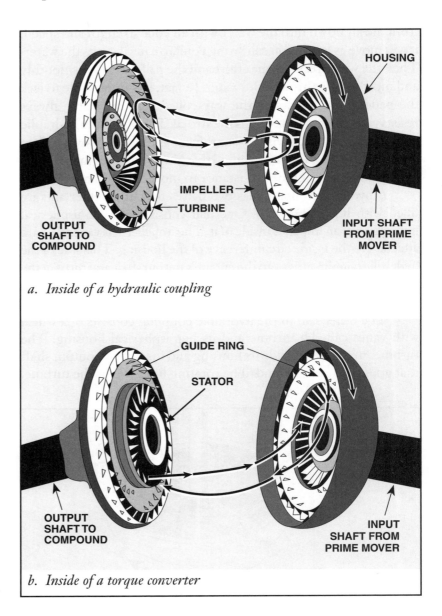

Figure 12. Hydraulic coupling and torque converter

a. *Inside of a hydraulic coupling*

b. *Inside of a torque converter*

The output shafts coming out of each of the hydraulic couplings are the input shafts for the compound. In other words, the shafts transmit power *out* of the prime mover and *into* the compound. Chains link sprockets on the ends of these shafts together (see fig. 9, "rotating shafts"). Each set of chains and sprockets from one shaft to another is a drive. The left side of figure 13 shows a view of these drives from above. The right side shows the selective transmission (drawworks). The shaft that powers the mud pumps is on the left side of the compound, and the output shaft to the selective transmission is in the center. The three engines, not shown, are at the bottom.

Design of the Compound

The term *sprocket* has two meanings. First, it refers to a cylinder with teeth on the outside of it that fits over the end of a shaft. These teeth are shaped and spaced to mesh with a chain (fig. 14). It can also refer to one of the teeth.

Chain-and-Sprocket Drives

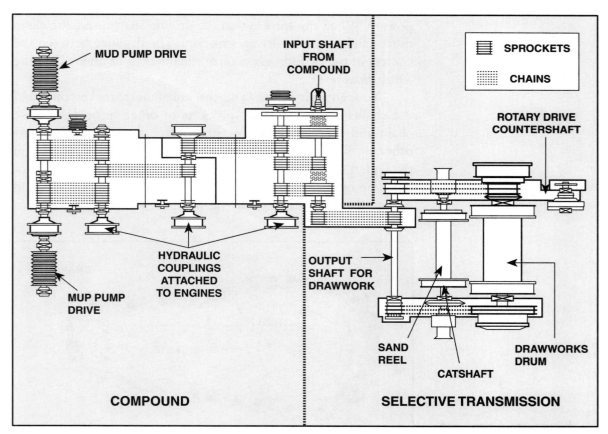

Figure 13. View of a compound for three engines from above

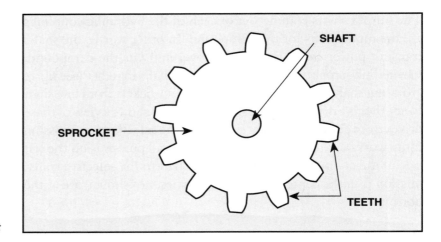

Figure 14. Sprocket

Sprockets work in pairs—two sprockets mesh with one chain (a chain-and-sprocket drive; fig. 15). Each chain-and-sprocket drive consists of a loop of chain that goes around two shafts with sprockets fitted on the shafts. One shaft is the driving shaft (it provides the power) and the other is the driven shaft (it receives the power). When the sprocket on the driving shaft turns, the chain moves and turns the driven sprocket. The distance between the centers of the two sprockets or the shafts they fit on is called the *center distance*.

You can see in figure 13 that each shaft has sprockets on it, and the chains connect certain sprockets to other sprockets. These chain-and-sprocket drives transmit power from one shaft to another, until the power is exactly where it needs to be. The shaft on the far left that goes to the mud pumps is the only drive that is not made up of chains and sprockets. Instead it uses large belts and pulleys that work exactly like the chain-and-sprocket drive.

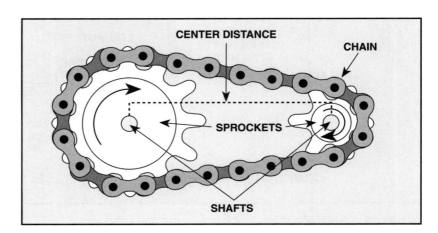

Figure 15. Chain-and-sprocket drive

Sprocket Ratio

The two sprockets in a drive can be the same size or different sizes. Same-sized sprockets transmit power without changing its speed or force. Drives with different-sized sprockets do change the speed and force (see fig. 15).

Imagine two sprockets, a small one with 6 teeth and a larger one with 12 teeth. When the smaller sprocket rotates one full turn, its 12 teeth engage with 12 links on the chain. Those ten links then engage with ten teeth on the larger sprocket, but it only turns halfway around. It takes two revolutions of the smaller sprocket to rotate the larger sprocket once. So the larger sprocket moves twice as slowly as the smaller sprocket, and the smaller sprocket moves twice as fast as the larger sprocket. Either one can be the driver, but the larger sprocket always moves more slowly.

The rule to remember is that when a large sprocket drives a smaller sprocket, the speed *increases*. When the smaller sprocket is the driver, the speed *decreases*.

Recall that a machine that changes either force or speed always changes the other in the opposite way. Since a chain-and-sprocket drive with different-sized sprockets changes the speed, it also changes the force. For example, a bicycle has a chain-and-sprocket drive between the shaft to the pedals and the axle of the back wheel. The pedal sprocket is the driving sprocket, and the chain connects it to the driven sprocket on the wheel. If the sprockets were the same size, when you pedaled one revolution with your feet, the back wheel would also turn one revolution. This would be easy pedaling (it would not require a lot of force), but you would move ahead very slowly. However, because the pedal sprocket is larger than the wheel sprocket, you have to pedal only one revolution to turn the wheel around several times, and so you move faster. But it takes more muscle strength to supply enough force to increase the speed.

When the pedal sprocket is close to the same size as the wheel sprocket, it is easy to start moving the bicycle, but hard to go very fast. When you switch speeds to a larger sprocket, on a ten-speed bike, for instance, it is harder to pedal at first but you can eventually go much faster. The greater the difference in size between the two sprockets, the greater the difference in speed and force between them.

The amount of difference in size between the two sprockets is the *sprocket ratio*. Having various combinations of sprocket ratios in the transmission allows the driller to vary the speed and amount of force to the hoist.

Multistrand Drives

Most chain-and-sprocket drives in both the compound and the selective transmission are *multistrand* drives (fig. 16). This means that each shaft actually has several sprockets, each with its own chain, or strand. The chains are connected to each other, so all the sprockets and chains on one shaft move at the same time and still make up only one drive. A drive with several strands is stronger than a drive with only one chain.

Figure 16. Multistrand drives

The compound also has a clutch between each chain-and-sprocket drive and prime mover. Each clutch is a mechanical on-off switch that allows the driller to select how many engines to compound for a particular operation (see Clutches).

A housing that is essentially a big steel box covers the compound. Like the guards on the drawworks, the compound's housing protects workers from the machinery and also confines a lubricating oil spray inside.

Other Components of the Compound

Several types of compounds are available. Small rigs generally use a one-piece, nonseparable compound. This type is small enough and rigid enough to move with the drawworks, and its chains do not have to be taken off for moves.

Larger rigs use compounds that can separate into sections for moving. The big disadvantage of this type of transmission is that the crew must take off the chains to move it. Some sectional compounds have hinges between sections so that they can swing open and fold back. Their chains can stay connected for moves, but this compound takes up more room than the type that separates completely.

V-belt compounds use belts, or *power bands*, instead of chains. Only the smallest rigs use this type, because belts are not strong enough for the heavy work that larger rigs do. On large mechanical rigs, however, power bands often drive the mud pumps. Since the pumps are usually a good distance from the compound, V-belts make it easier to connect the pumps to the compound. A distance of 15 feet (4½ metres) would require a 47-foot (14-metre) chain, which is very heavy and awkward to work with. A belt that long is much less trouble. The disadvantages of belts are that they take up more space than chains (because they are wider and because it takes more of them to transmit the same torque) and the tension has to be adjusted frequently. Their main advantage is that they do not have to be lubricated.

Types of Compounds

Selective Transmission

The second type of transmission in the hoisting system is the selective transmission. It has several separate drives and clutches that transmit power to the hoisting drum, the catshaft, and, usually, the rotary table. The purpose of the selective transmission is, first, to select where the power will go. Second, it also changes the speed and force relationships, so the driller can determine how much power goes to each component.

Types of Selective Transmissions

A selective transmission system may use either chain-and-sprocket drives, like the compound, or gears. The driller controls these with levers at the console.

Gear Transmissions

A gear is a metal wheel with teeth around the outside circumference. Gears, like sprockets, work in pairs, but without the chain—the teeth of two gears mesh directly with one another (fig. 17). Each gear fits onto a shaft that rotates. As in chain-and-sprocket drives, one of the shafts is the driving shaft (called the pinion) and the other is the driven shaft. Also like chain-and-sprocket drives, when the two gears are different sizes, they change the speed and the force from one shaft to another. The only difference is that gears also change the direction of rotation from one shaft to the next.

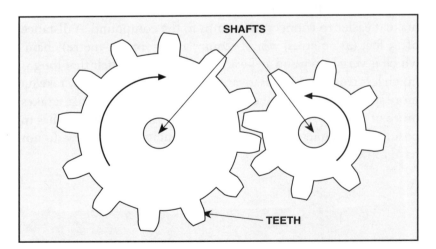

Figure 17. Gears

Small drawworks for lighter-duty hoisting often use gears in the selective transmission because they take up less room than chain-and-sprocket drives. Larger drawworks cannot use gears because the teeth would break under the heavier loads.

Gear transmissions generally have a torque converter instead of a clutch at the connection to the compound. As described above, a torque converter smooths the power flow, so it reduces the shock to the gear teeth, which prevents them from breaking.

Larger mechanical-drive drawworks also have a pair of gears between the compound and the selective transmission on the input shaft (see fig. 18a). Because gears reverse the direction of movement, they allow the driller to reverse the direction of rotation of the drum. The only time the driller does this is when the crew is servicing the drawworks, not when hoisting.

Chain-and-Sprocket Transmissions

Most rigs use a selective transmission made up primarily of chain-and-sprocket drives. The power flow diagrams in figure 18a and b show the drawworks from above. Selective transmissions on electric and mechanical drives are identical below the output shaft.

Between the input shaft coming from the compound and the output shaft (*jackshaft*) on a mechanically driven drawworks (fig. 18a) are three two-strand chain-and-sprocket drives (nos. 1, 2, 3) and a pair of gears for reversing (no. 4). Each of these chain-and-sprocket drives has a different sprocket ratio to provide three speeds to the high drum drive and three to the low drum drive.

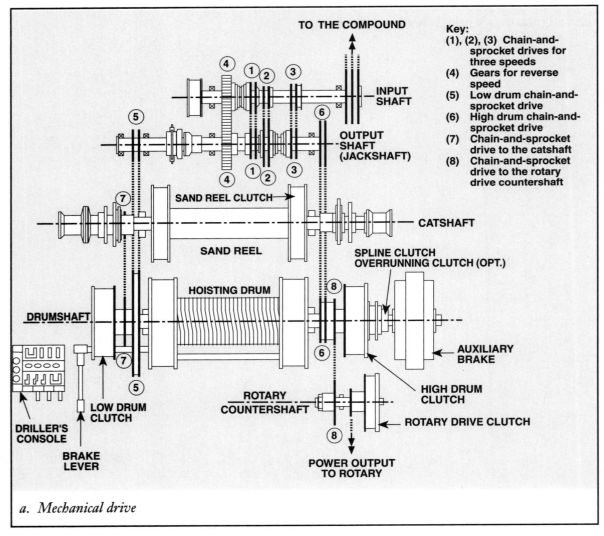

Key:
(1), (2), (3) Chain-and-sprocket drives for three speeds
(4) Gears for reverse speed
(5) Low drum chain-and-sprocket drive
(6) High drum chain-and-sprocket drive
(7) Chain-and-sprocket drive to the catshaft
(8) Chain-and-sprocket drive to the rotary drive countershaft

a. Mechanical drive

Figure 18. Power flow diagram of the selective transmission

Manufacturers also sell drawworks with two or four drives here. The low drum chain-and-sprocket drive (no. 5) is on the driller's side of the output shaft, and the high drum drive (no. 6) is on the off-driller's side. They connect the output shaft to the drum. By engaging and disengaging different clutches to all of these chain-and-sprocket drives, the driller running this drawworks has six hoisting speeds.

The electrically driven drawworks (fig. 18b) has two motors that power the input shaft and no reversing gears.

A single-strand drive on the driller's side (no. 7) connects the drumshaft to the catshaft. A final single-strand drive (no. 8) connects the drumshaft to the rotary drive countershaft on the rotary side.

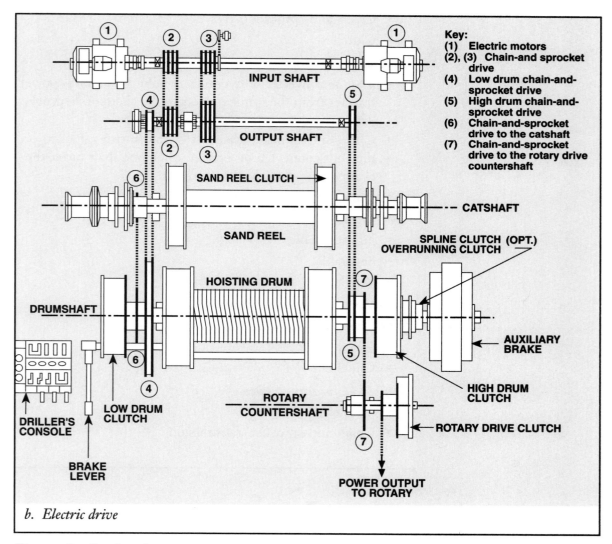

Key:
(1) Electric motors
(2), (3) Chain-and sprocket drive
(4) Low drum chain-and-sprocket drive
(5) High drum chain-and-sprocket drive
(6) Chain-and-sprocket drive to the catshaft
(7) Chain-and-sprocket drive to the rotary drive countershaft

b. Electric drive

Figure 18—Continued

To summarize—

Types of transmissions

- Compounding transmission, or compound—in a mechanical-drive rig only, transmits power from the prime movers to the drawworks
- Selective transmission—in all rigs, transmits power from the compound or electric motors to the drum, catshaft, and rotary table countershaft

Components of the compound

- Torque converter or hydraulic coupling
- Chain-and-sprocket drives
- Clutches

Job of each component

- Torque converter or hydraulic coupling—smooths power transfer from the prime movers to the chain-and-sprocket drives
- Chain-and-sprocket drives—two shafts with sprockets on the ends connected by a chain; when one shaft turns, the chain causes the other shaft to turn
- Clutches—engage or disengage the drives

Types of compounds

- Nonseparable
- Separable
- V-belt

Components of the selective transmission

- Chain-and-sprocket drives
- Gears
- Clutches

Types of selective transmissions

- Gear transmissions
- Chain-and-sprocket transmissions

Everything in the following sections about chain-and-sprocket drives applies to both the compound and the selective transmission.

The chain used for rig transmissions is *roller chain*. Roller chain on rigs looks and works like a bicycle or motorcycle chain, but it is much larger. The links of roller chain on rig transmissions may be as long as 3 inches (7½ centimetres) and they are very strong. Roller chain has two types of links that alternate: roller links and pin links (fig. 19).

Construction of Chains and Sprockets

Chain Construction

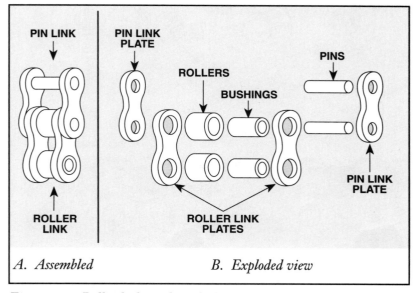

A. *Assembled* B. *Exploded view*

Figure 19. Roller links and pin links

Roller links consist of two bushings whose ends fit into *roller link plates* (also called *sidebars*) and two rollers that fit over the bushings. The bushings cannot rotate; they are fixed. The rollers are the parts that engage with the teeth of the sprockets. Because they roll when the chain meets the sprocket, they also act as shock absorbers and reduce the effects of the impact. Imagine skating on roller skates. As long as the wheels turn freely, you can move ahead smoothly. If the wheels of one skate suddenly lock up, you will probably fall flat on your face. When the roller wheels are working properly, they absorb the shock from the impact of your foot hitting the sidewalk and convert it into a rolling movement. When they don't, your body has to absorb the forward motion.

Pin links have two pins whose ends are immovably fixed into pin link plates. The pins fit inside the bushings of the roller links. As the chain goes around a sprocket, it bends only between the pin and the bushing. The pins are riveted to the link plate on one side and either riveted or fixed with cotters to the other pin link plate (fig. 20). A cotter is a spring steel wire that fits into an eye on the pin outside the plate. Chain in a rig transmission is usually cottered chain.

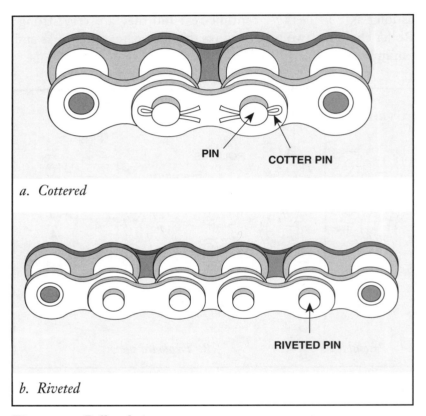

a. Cottered

b. Riveted

Figure 20. Roller chain

In order to make a continuous loop of chain, a third type of link, a *connector link*, connects the two ends of a chain. The connector link is a pin link with either a spring clip (fig. 21a) or a cotter (fig. 21b) to hold the pins. The cottered type looks, and sometimes is, the same as the cottered pin link. So a cottered chain is actually made up entirely of connector links and roller links.

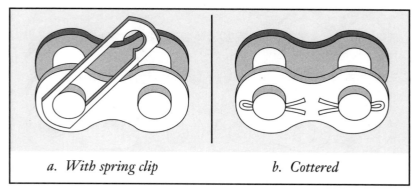

| a. With spring clip | b. Cottered |

Figure 21.   Connector links

A fourth type of link is the *offset link* (fig. 22). An offset link is a combination of roller link and pin link. Its link plates have a bend in them so that one end has a pin and the other end has a roller. A chain will need an offset link to adjust its length by one pitch, or one link (see "Chain Dimensions" on next page).

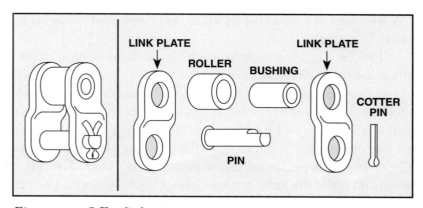

Figure 22.   Offset link

Roller chain with a single strand looks like bicycle chain. Drives in the drawworks and compound may have up to four strands (fig. 23). The pins in a multistrand chain span the whole width of the chain, through all strands, so that the strands move together as one chain.

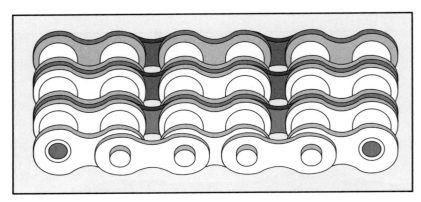

Figure 23. Multistrand chain

Chain Dimensions

The *pitch* of a roller chain is the distance in inches (millimetres) between the centers of the bushings (fig. 24). People in the drilling business use the word "pitch" to talk about the length of a chain. For example, they would say that a chain has 100 pitches, which is the same as having 100 links. One set of a roller link and a pin link has two pitches. So a chain normally has an even number of pitches because two roller links or two pin links cannot connect together. The only way to have a chain with an odd number of pitches is to use an offset link.

Pitch is the dimension from which the manufacturer determines the other dimensions of the links.

The *roller width*, or chain width, is the width of the rollers, which is the distance between the inside faces of the roller link plates. It is not the width of the chain between the outside faces of the link plates nor is it the width of the pins (they are both wider). The *roller diameter* is the outside diameter of the roller. The roller width and roller diameter are both about ⅝ of the pitch. The *pin diameter* is about 5/16 of the pitch. The thickness of the link plate is about ⅛ of the pitch, or more for heavy-duty chain.

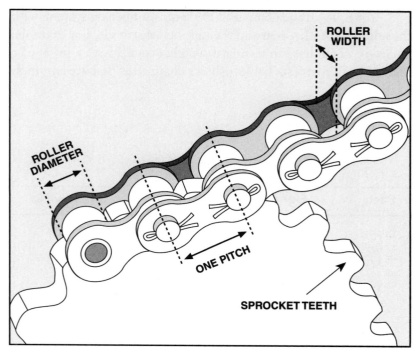

Figure 24.   Dimensions of a link

Manufacturers label every chain with a number that ANSI (American National Standards Institute) has defined. Single-strand chain is identified with a two- or three-digit number. The first one or two digits of this number tell the pitch in eighths of an inch. A chain numbered 80 has a pitch of 8 eighths, or 1 inch (about 25 millimetres). A number 160 chain has a pitch of 16 eighths, or 2 inches (about 50 millimetres). The last digit identifies the relative dimensions of the parts of the links, as described above. This digit is 0 for standard-weight chain, 1 for lightweight chain, or 5 for rollerless chain. An H after the number means that the chain is a heavy-duty chain. Heavy-duty chain has link plates that are as thick as the next larger size standard chain. Only a heavy-duty chain can have an H after its number.

Multiple-strand chain uses the same numbering system, with the addition of a hyphen and the number of strands. For example, an 80–10 chain has ten standard-weight strands with a pitch of 1 inch (25 millimetres). Table 1 shows chain sizes that are common in the oilfield.

Table 1
Common Chain Sizes

Pitch	Eighths	Std. No.	Heavy No.
0.25	2	25-rollerless	None
0.375	3	35-rollerless	None
0.50	4	41-light duty	None
0.50	4	40	None
0.625	5	50	None
0.75	6	60	60H
1.00	8	80	80H
1.25	10	100	100H
1.50	12	120	120H
1.75	14	140	140H
2.00	16	160	160H
2.25	18	180	180H
2.50	20	200	200H
3.00	24	240	240H

Source: Diamond Chain Company.

Chain-and-sprocket life are affected by three limits to the strength of the metal they are made of: ultimate strength, fatigue limit, and galling limit. *Ultimate strength* refers to how much *one-time* stress the chain or the sprocket can withstand before breaking. For example, the ultimate strength of a pencil is exceeded when someone snaps it in two. One severe shock will break the pins when a shock exceeds the ultimate strength of the chain (fig. 25). The strength of the chain determines how much weight it can hold without breaking, so the manufacturer rates the ultimate strength value of the chain at a certain weight. The driller uses this rating to determine how much weight a chain can hold when it is not moving. Most chain manufacturers recommend using only 40% of the chain's ultimate strength. For example, if the chain is rated at 10 tons, the driller should not allow it to hold more than 40% of that, or 4 tons, when the chain is stopped.

Fatigue refers to the tendency of metal to break under *repeated* stress. For example, a person can break the ring tab off a can of soda by bending it back and forth. A chain wears out when its link plates reach their fatigue limit and break (fig. 26). The load, or weight, on the chain and the number of times it revolves around the sprockets determine how fast it will wear out because of fatigue.

Design Limits of Metals in Chain and Sprockets

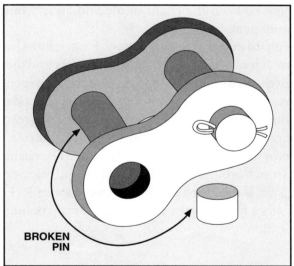

Figure 25. Broken pin

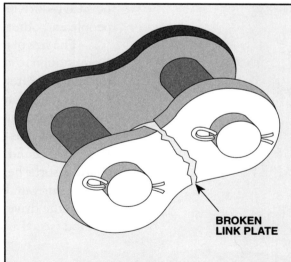

Figure 26. Broken link plate

37

Galling is wear caused by friction. The metal of the pins and the bushings gradually wears off as these two parts rub against each other. Galling generally occurs because of poor lubrication in chain that moves slowly. But it can occur even in well-lubricated, fast-moving chain because the metal-to-metal contact can cause the metal to reach its galling limit. The galling limit just means that the metal wears even though it is lubricated because of a heavy load or a high speed. The pins and the bushings are the parts that break due to galling.

Chain Guards

All chain-and-sprocket drives have a *chain case*, or *chain guard*, to protect workers from the moving drive. It also protects the drive from dirt and is part of the cooling and lubrication systems. Some enclose only one drive, such as the high drum drive (fig. 27a), and some enclose several drives (fig. 27b). A chain guard may be made of sheet metal or of plate metal, which is much thicker than sheet metal (perhaps ¾ inch, or 19 millimetres). In addition to surrounding the drive, these heavier guards also support the shafts on bearings. Both types have access panels for inspection and maintenance.

When a chain in a high-speed drive breaks, it goes flying at perhaps 4,000 feet (1,200 metres) per minute. The chain case must be heavy-duty enough to keep the chain inside and away from people and other equipment.

The size of the guard depends on two factors. First, it must be large enough to allow at least 3 inches (7.6 centimetres) around the top and the bottom of the chain and ¾ inch (19 millimetres) on each side. As the chain wears, it starts to sag on the slack side (the side moving away from the driving shaft). The sag can become so great that the chain hits the case, which damages both the chain and the case.

The second factor in guard size is cooling capacity. The chain guard absorbs heat from the oil lubricating the chain and releases it to the surrounding air. If the guard gets hotter than about 180°F (82°C), the drive needs a larger guard or a supplemental cooling method (see "Lubrication").

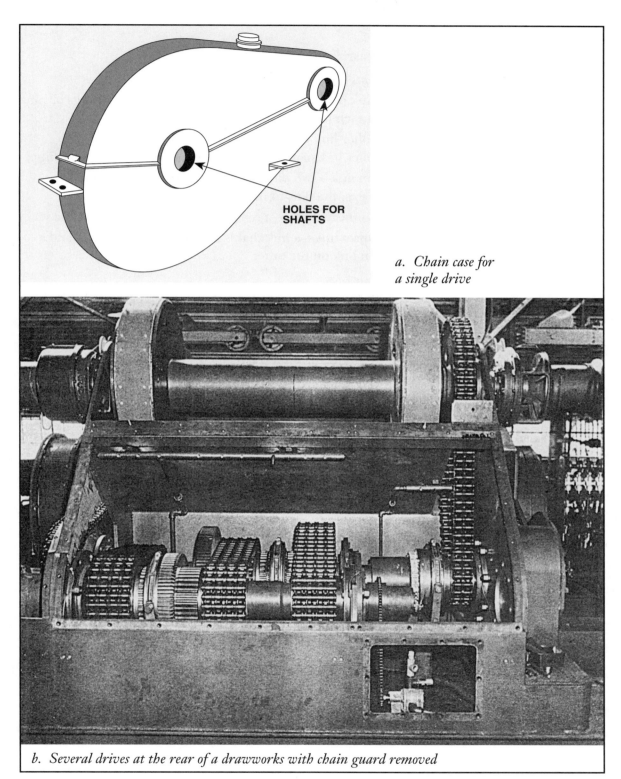

HOLES FOR SHAFTS

a. Chain case for a single drive

b. Several drives at the rear of a drawworks with chain guard removed

Figure 27. Chain case and chain guard

To summarize—

Components of a roller chain

- Link plates—the side plates that hold the bushings or pins of each link
- Roller link—two bushings whose ends are fixed into roller link plates and two rollers that fit over the bushings
- Pin link—two pins whose ends are fixed into pin link plates; the pins fit inside the bushings of roller links
- Connector link—a pin link with a removable link plate
- Offset link—a link that is a roller link on one end and a pin link on the other

Chain dimensions

- Pitch—the distance between two bushings of a link
- Roller width—the width between the insides of the link plates
- Roller diameter—the diameter of the rollers
- Pin diameter— the diameter of the pins

Types of metal strength limits in roller chain

- Ultimate strength—how much one-time stress the metal can withstand before breaking
- Fatigue—the tendency of a metal to break under repeated stress
- Galling—wear due to friction

Purposes of chain guards

- To protect workers from the moving drive
- To protect the drive from dirt and damage
- To hold in oil from the lubrication system
- To absorb heat from lubricating oil and to release that heat to the air

Keep new chain in its box until ready to use it to preserve the factory lubrication and to prevent contamination by dirt and debris. If a new chain is not the correct length to fit on the drive, shorten or lengthen it by removing or adding links. Avoid changing the pitch by one if possible. Although offset links are commonly available and are the only way to add a single pitch, as they wear, the bend in the link plates tends to straighten. This causes the chain to get longer and defeats the original purpose of using offset links. Also, they can be much weaker than the rest of the chain.

Chain comes from the manufacturer unconnected, not as a loop. Fit it around the sprockets and bring the two ends together, using the sprocket teeth to hold the ends in position. Insert the pins of the connecting link through the bushings to make the loop. Push the link plate over the pins. Then fasten the pins with the spring clip or cotters. Last, press the pins back until the fasteners are snug against the outside of the link plate. This ensures that the connector link's pins line up with the rest of the pins and so have the same clearance. It also helps the joint to flex as it should.

It is best to use chain from one manufacturer on any particular drive, or the drive may run rough. However, if it is necessary to replace one brand with another, the specifications and dimensions of the new chain must match those of the original chain. That is, the strength and type of material must be the same, and also the pitch, the roller diameter, and the roller width. Otherwise the new chain will not fit on the old sprockets. Some chains can actually intercouple with other brands or types, so that a floorhand could replace just a section of a chain. Manufacturers do not recommend doing this except when unavoidable. In an emergency, it is possible to repair a chain with parts (pins, link plates, etc.) from a different type or brand of chain. However, the dimensions and design are almost never identical, so the repaired section does not fit properly or last very long. If it is impossible to avoid doing this as a temporary fix, replace the chain as soon as possible.

Installing Chain

Chain Tension

After installing new chain, adjust the tension (fig. 28). First, tighten one span of the chain by rotating the sprockets in opposite directions. A span is the section on one side of the sprockets. Then measure the sag, or *catenary*, of the slack span by the method given in the next paragraph. The catenary should be 4 to 6% for drives that are horizontal up to an incline of 45°, or 2% to 3% for drives that are from 45° to vertical.

To calculate the maximum acceptable catenary, lay a straight-edge across the tops of the sprockets. Measure the distance from the middle of the span to the point where it touches the sprocket (say this is 10 feet, or 3 metres). Multiply this number by the percentage of acceptable catenary, say 5% for a horizontal drive. This gives a number of 0 feet × .05 = .5 foot, or 6 inches (3 metres × .05 = 1.5 metres, or 15 centimetres).

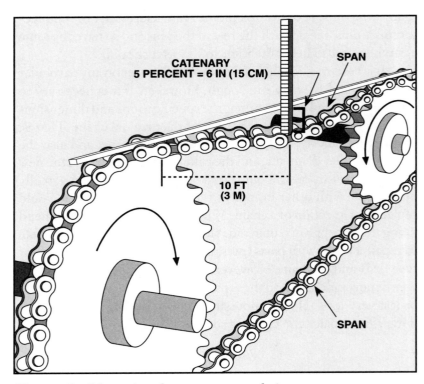

Figure 28. Measuring the catenary on a chain

Then measure the distance from the middle of the span to the bottom of the straightedge. This distance is the actual catenary and should be between 4% and 6%—6 inches (15 centimetres). If it is more than 6%, the chain is too loose; if it is less than 4%, the chain is too tight.

The chain will settle in after a few hours of running and the crew should recheck the catenary then. Never try to check the tension while the chain is moving.

Setting the proper chain tension is important for several reasons. If the tension is too tight, the pins and the bushings may gall, or wear because of friction, because the lubricant cannot get into the spaces between them. In extreme cases, the chain can be so tight that the pitch in the chain becomes longer. Then the chain will not fit the sprockets properly, causing the link plates to break.

If, on the other hand, the chain is too loose, there will not be enough tension to pull the links away from the driving sprocket. The chain will follow the sprocket up as it turns and then jerk away (fig. 29). This will eventually damage the tips of the sprocket teeth, and the chain rollers will crack because of the shock of the constant jerking. Loose chain also produces vibration that may wear out the pins and the bushings. Finally, the chain may strike the chain case, damaging both chain and casing.

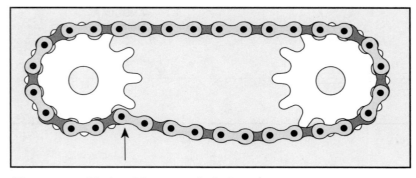

Figure 29. Chain with too much slack

Chain Alignment   The consequences of not aligning chain-and-sprocket drives properly is one of the reasons electric drives are easier to maintain. The manufacturer originally aligns chain-and-sprocket drives, but the crew must check them when rigging up at a new site and if a wear problem indicates that they are out of alignment. Two types of misalignment, angular and offset, affect the life of the chain.

In *angular misalignment*, the shafts are not parallel to each other (they form an angle) in either the horizontal or the vertical plane (fig. 30). This pulls the link plates on one side tighter than those on the other side; thus, one side of the chain and sprockets wears faster than the other. Link plates on only one side of the chain break because of fatigue, which is a way of discovering angular misalignment.

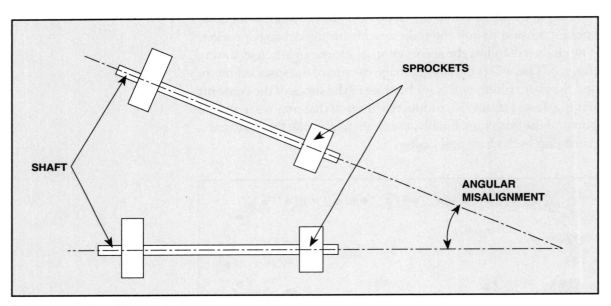

Figure 30. Angular misalignment

In *offset misalignment*, the ends of the shafts, and therefore the sprockets, are not in line with each other (fig. 31a). Some sprockets can slide on the shaft, so it is possible for them to be misaligned even if the shafts are not (fig. 31b). Slight offset will not cause a problem, but the amount depends on the distance between sprockets, chain pitch, and number of chain strands. Offset misalignment alternately stresses the link plates on one side of the chain and then those on the other side. One sign of offset misalignment is wear on one side of the driver sprocket and on the opposite side of the driven sprocket. Another sign is that the outer link plates on both sides of a multistrand chain will break because of fatigue, but not the inner link plates.

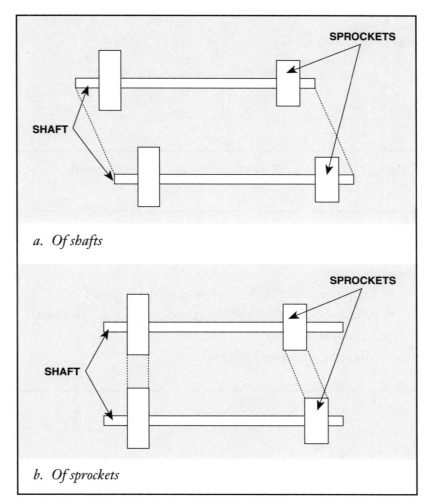

Figure 31.   *Offset misalignment*

Combined misalignment may be the result of combined angular and offset misalignment, or it may be the result of two shafts that are not level with each other (fig. 32). Since either angular or offset misalignment can break the chain, it is wise to check for both and also for levelness of the shafts. Do this by laying a straightedge across the two shafts.

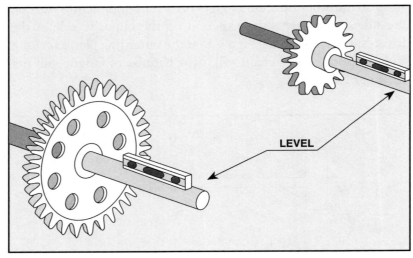

Figure 32. Checking for misalignment and levelness of shafts

To summarize—

Installing chain

- Keep new chain in the box until ready to install it
- Adjust the length of the chain and connect it into a loop
- Adjust the tension
- Align the shafts and sprockets

Each joint in the roller chain of every chain-and-sprocket drive in the compound and in the selective transmission is a kind of bearing, and bearings cannot work without lubrication. A lubricant also smooths the engagement of the chain with the sprockets, dissipates heat, flushes away debris, and retards rust. Oil is the lubricant in chain-and-sprocket drives.

Oil must enter the chain between each roller link plate and pin link plate, the only opening to the pin and inside of the roller (fig. 33). Then it must seep into the spaces between the pin and bushing and between the bushing and roller. There are three methods for lubricating roller chain drives. The method of applying the oil depends on the speed of the drive (see table 2).

Lubrication of Chain-and-Sprocket Drives

Methods of Lubrication

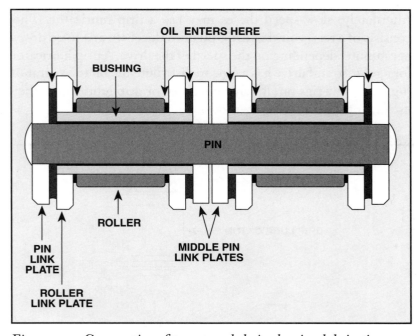

Figure 33. Cross section of two-strand chain showing lubrication

Manual or drip lubrication

On slow-speed drives, a floorhand can pour oil directly on the chain with a spout can or paint the chain with an oiled brush, once every 8 hours, while the drive is stopped. It can be dangerous to try this while the chain is moving, as the term "slow" is relative: a slow-speed drive may move up to 300 feet (about 90 metres) per minute.

Table 2
Maximum Chain Speeds for Different Lubrication Methods
(Feet/Metres per Minute)

Chain no.	35	40	50	60	80	00	20	40	60	200
Manual or drip	350/107	300/91	250/76	215/66	165/56	145/44	125/38	110/34	100/30	80/24
Bath or disk	2,650/808	2,200/671	1,900/579	1,750/533	1,475/450	1,250/38	1,70/357	1,050/320	1,000/300	865/264
Pressure	Use for speeds higher than bath or disk limits									

Source: Diamond Chain Company. Used with permission.

Alternately, slow-speed drives may use a drip lubricator. This consists of a reservoir that drips oil onto the chain at 4 to 20 drops per minute, depending on the speed of the drive. A drip lubricator for a multistrand drive has a pipe to distribute the oil to all strands (fig. 34). New rigs rarely rely on manual or drip lubrication; they use one of the following methods.

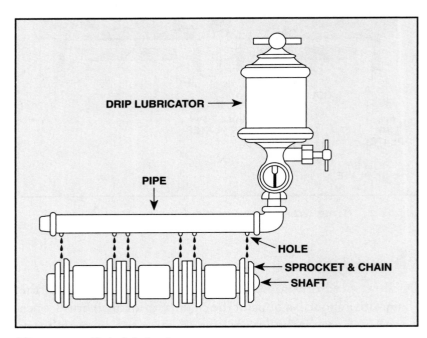

Figure 34. Drip lubrication

Bath or Disk Lubrication

Higher-speed drives, up to about 2,650 feet (about 800 metres) per minute, may use oil bath or disk lubrication. In the oil bath method, a portion of the chain passes through an oil bath (sump), which coats all of the chain on each revolution to lubricate it (fig. 35).

Another version of oil bath lubrication is slinger disk lubrication. Here, the chain itself does not pass through the sump, but one or two disks rotating with the sprocket pick up oil from the sump and sling it against a plate, which then feeds it to the top of the lower span (fig. 36).

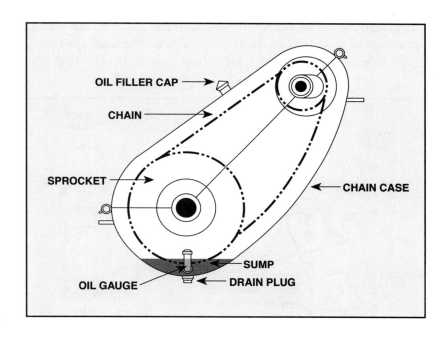

Figure 35. Oil bath lubrication

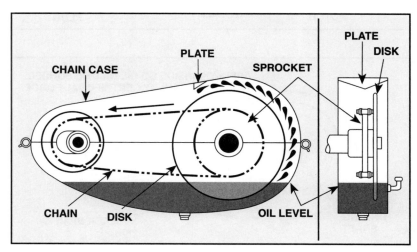

Figure 36. Slinger disk lubrication

Pressure Lubrication

Drives with speeds higher than 2,650 feet (800 metres) per minute are moving so fast that to run them through an oil bath would just spray oil away from the chain like a boat speeding through water. Oil would never reach the tiny spaces between the bushings and the pins. So high-speed drives use *forced* or *pressure lubrication*. Heavily loaded drives also use pressure lubrication.

Pressure lubrication uses a pump to force oil continuously through pipes with holes in them, through nozzles, or through holes in the sprockets (fig. 37). In any of these, oil sprays out directly onto the chain with pressure. On some rigs, the oil sprays onto the outside of the chain, and on others onto the inside of the chain. Spraying the inside of the chain, preferably on the slack span (that is, the side that is not driven), is probably more effective (fig. 38).

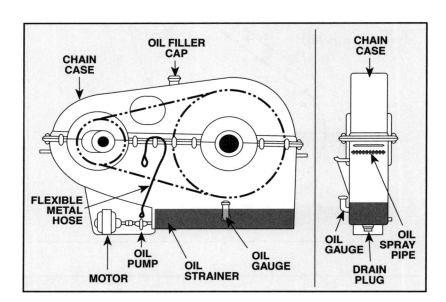

Figure 37. Pressure lubrication

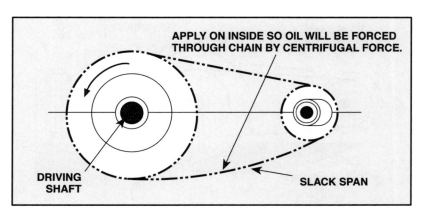

Figure 38. Best location to apply lubrication

The driver is not stretching the slack span, so the spaces inside and between pins, bushings, and rollers are a little larger. When the pressure spray is on the inside of the chain, centrifugal force pushes the oil through the chain into the area that most needs it.

The guards around drives with pressure lubrication must seal tightly in order to keep the oil inside. This prevents it from polluting the environment.

Friction between the chain and the sprocket teeth creates heat. A more important source of heat in bath lubrication, however, is oil shear, or friction from the chain passing through the oil in the sump. The higher the oil level in the sump, the hotter the oil gets. When it is too hot, it foams. The solution to foaming oil is not to add more oil to the sump, but to check to see if there is too much already. The oil level should be just high enough to cover the chain as it passes through it.

Cooling the Oil

Some drives can stay cool enough by releasing heat to the chain case and then to the air, but high-speed or heavily loaded drives using pressure lubrication need a better cooling method. They use oil-to-water heat exchangers. In this case, the heat exchanger is a pipe inside the chain case that has cool water flowing through it. Heat from the chain and the oil spray heats the water, which then flows outside and cools before recirculating through the chain case.

In addition to lubricating moving parts, the oil has another job—to clean the chain. As the chain wears, tiny particles of metal wear off and contaminate the oil. These particles and any grit in the oil damage the metal parts that the oil is supposed to be protecting. If you have ever been outdoors in a sandstorm, you know how powerful tiny particles of sand can be when they are moving fast. Change the oil whenever it looks dirty, at least once every 500 hours of operation, or every 200 hours under severe conditions. After installing a new chain, change the oil after the chain has run 50 hours. A new chain flakes off metal particles relatively quickly as it breaks in.

Maintaining the Oil

The crew should check the oil level in the sump every 8 hours of running time and add oil as needed. At the same time, crew members should check for leaks, foaming, and overheating. If the oil appears yellow and foamy, change it.

Viscosity

Be sure to use a brand-name, nondetergent oil with the viscosity that the chain manufacturer recommends. Viscosity is a measure of how thick the oil is. Oil becomes more viscous in colder temperatures. Thus you should use an oil with lower viscosity in the winter than in the summer.

To summarize—

Methods of lubricating chain-and-sprocket drives

- Manual—a floorhand brushes or pours oil onto the chain
- Drip—a pipe with holes in it drips oil onto the chain
- Bath—part of the chain passes through an oil bath (sump)
- Disk—a rotating disk passes through a sump and slings oil onto the chain
- Pressure—a pump forces oil through holes or nozzles onto the chain

Methods of cooling the oil

- By releasing heat to the chain guard
- With a heat exchanger

When to change the oil

- 50 hours after installing new chain
- On a regular schedule or when it is dirty
- If it is foamy and yellow

To maintain the selective transmission and the compound, inspect them every tour. At each inspection, check the following parts and service or repair them if necessary.

Maintenance of Chain-and-Sprocket Drives

For a manual lubrication system, clean the chain with kerosene or a nonflammable solvent if it is dirty. Then brush or pour on the correct type of oil.

For drip lubrication, check oil level and flow rate, and be sure the oil is dripping on the correct part of the chain.

For oil bath, slinger disk, or pressure lubrication, be sure all outlet holes are clear and that the oil is hitting the right part of the chain. Check the oil level. Look for oil leaks. Leaking oil pollutes the environment.

Lubrication System

Inspect the chain for cracked, broken, deformed, or corroded parts and for tight joints or turned pins. If you find any, first find and correct the cause of the damage. Then replace the entire chain. Even if the rest of the chain looks all right, it probably has been damaged. If 5% or more of the thickness of the chain's link plates is worn away or if there are sharp gouges in the surface of the link plates, replace the chain immediately. A tool called a micrometer, or "mike," measures accurately to thousandths of an inch (.025 millimetre). Use it to measure the worn thickness and compare it to the original new thickness of the link plate.

Chains and Sprockets

In most roller chain drives, the chain is considered to be worn out when it has stretched by 3%. With this amount of stretching, or elongation, the chain no longer engages properly with the sprockets, and this can damage the teeth or break the chain. In some drives, the allowable wear is much less.

Chain Wear

To determine the wear elongation, use a table like table 3. First tighten one span of the worn chain by rotating the sprockets in opposite directions. Then find the ANSI number of the chain you are testing in table 3. Move across to the column that gives the number of pitches. Measure this many pitches on a section of the tight chain (fig. 39). For example, if the chain is a #25, measure 48 pitches. Then compare this measurement to the number in the last column. In this example, the length of the chain when new was 12 inches, or 305 millimetres, and the maximum stretched length is 12.375 (= 12⅜) inches, or 314 millimetres. If the measurement of the worn chain is equal to or greater than the number in the last column, replace the chain.

Table 3
Elongation Limits for Chain

Pitch No.	Chain Pitch		Measured Length				
	in.	mm	No. of Pitches	Nominal		At 3% Elongation	
25	.250	6.35	48	12.00	305	12.375	314
35	.375	9.52	32	12.00	305	12.375	314
41	.500	12.70	24	12.00	305	12.375	314
40	.500	12.70	24	12.00	305	12.375	314
50	.625	15.88	20	12.50	318	12.875	327
60	.750	19.05	16	12.00	305	12.375	314
80	1.000	25.40	12	12.00	305	12.375	314
100	1.250	31.75	20	25.00	635	25.750	654
120	1.500	38.10	16	24.00	610	24.719	628
140	1.750	44.45	14	24.50	622	25.250	641
160	2.000	50.80	12	24.00	610	24.719	628
180	2.250	57.15	11	24.75	629	25.500	648
200	2.500	63.50	10	25.00	635	25.750	654
240	3.000	76.20	8	24.00	610	24.719	628

Source: IADC Drilling Manual, 11th ed., chap. G, sec. 2, table G2-4.

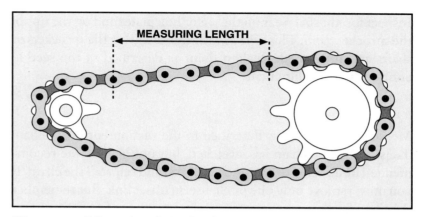

Figure 39. Measuring chain for elongation

It is harder to tell when a sprocket is worn out. Check for roughness or binding when a new chain engages or disengages with the sprocket. Inspect the teeth to see if they have become thin or hooked on the tips (fig. 40). Replace the sprockets if more than 10% of the tooth thickness has worn away.

Inspect the sprockets for chipped, broken, or deformed teeth. If you find any, correct the cause of the damage and replace the sprocket *and* chain. Sprockets are stronger and less sensitive to damage than chain, but running a worn chain on new sprockets can ruin the new sprockets very quickly. Likewise, do not run new chain on worn out sprockets.

Sprocket Wear

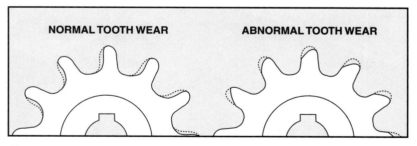

Figure 40. Worn sprockets

Misalignment   Inspect for unusual wear on the roller link plates and on the tips of the sprocket teeth. This type of wear indicates that the sprockets or shafts are misaligned. Realign them as described in the section entitled "Chain Alignment."

Proper Tension   Measure the tension as described in the section entitled "Chain Tension." If the chain has stretched, but not beyond the recommended limit, remove two pitches (links) and reinstall the chain. If you must remove only one pitch, use an offset link. But remember that it is better not to use offset links.

Chain Guards   If a chain case is bent, make sure that it still leaves enough clearance for the chain. Replace the guard if it is broken or damaged, especially near areas where it is mounted. Also replace it if its openings are enlarged.

Table 4 is a list of problems that inspection of the chain-and-sprocket drive may reveal, the possible causes, and the cures.

Table 4
Roller Chain Troubleshooting Guide

Condition/Symptom	Possible Cause	What to Do
Excessive noise	Chain striking on obstruction	Replace chain and elimate interference
	Loose casing or shaft mounts	Tighten bolts, realign drive, and retension chain
	Excess chain slack	Retension chain
	Excessive chain wear	Replace chain and sprockets and realign sprockets
	Sprocket misalignment	Replace chain and sprockets if indicated
	Inadequate lubrication	Realign sprockets; retension chain
	Chain pitch too large	Replace chain if indicated; reestablish proper lubrication
	Too few sprocket teeth	Redesign drive for smaller-pitch chain; check to see whether larger sprockets can be used
Chain climbs sprocket teeth	Excess chain slack	Retension chain
	Excessive chain wear	Replace and retension chain
	Excessive sprocket wear	Replace chain and sprocket and realign sprockets
	Extreme overload	Replace chain and eliminate cause of overload
Chain clings to sprocket	Excessive sprocket wear	Replace chain and sprockets and realign sprockets
	Sprocket misalignment	Replace chain and sprockets if indicated; realign sprockets.
Wear on inside of link plates and on one side of sprocket	Sprocket misalignment	Replace chain and sprockets if indicated; realign sprockets
Tight joints	Dirt or foreign material in chain joints	Clean and relubricate chain
	Inadequate lubrication	Replace chain and sprockets if indicated and realign sprockets
	Corrosion or rust	Replace chain and sprockets if indicated and realign sprockets
	Overloads bend pins or spreads roller link plates	Replace chain and eliminate cause of overload

Table 4, Continued
Roller Chain Troubleshooting Guide

Condition/Symptom	Possible Cause	What to Do
Turned pins	Inadequate lubrication	Replace chain and reestablish proper lubrication
	Overload	Replace chain and eliminate cause of overload
Enlarged holes	Overload	Replace chain and eliminate cause of overload
Cracked link plates (fatigue)	Loading above chain's dynamic capacity	Replace chain and eliminate cause of high loading, or redesign drive for larger chain
Broken pins	Extreme overload	Replace chain and eliminate cause of overload, or redesign drive for larger chain; replace sprockets if indicated
Broken link plates	Extreme overload	Replace chain and eliminate cause of overload, or redesign drive for larger chain; replace sprockets if indicated
Broken, cracked, or deformed rollers	High-speed impact, or sprockets too small	Replace chain. Possibly redesign drive for smaller-pitch chain or larger sprockets
	Chain riding high on sprocket teeth	Replace chain and readjust tension more often
Pin galling	Speed/load too high	Reduce speed or load. Possibly redesign drive for smaller-pitch chain
	Inadequate lubrication	Provide or reestablish proper lubrication

Table 4, Continued
Roller Chain Troubleshooting Guide

Condition/Symptom	Possible Cause	What to Do
Worn link plate contour	Chain dragging on case, guide, or obstruction	Replace chain when 5% of contour worn away; retension chain and eliminate interference
Battered link plate edges	Chain striking obstruction	Replace chain and eliminate interference
Missing parts	Missing at assembly	Replace chain
	Broken and lost	Replace chain
Rusted chain	Exposure to moisture	Replace chain and protect from moisture
	Water in lubricant	Replace chain and protect from moisture
	Inadequate lubrication	Provide or reestablish proper lubrication
Corroded or pitted chain	Exposure to corrosive environment	Replace chain and protect from hostile environment
Missing or broken cotters	Vibration	Replace chain; reduce vibration; use larger sprockets
	High speed	Replace chain; reduce speed; redesign drive to use smaller-pitch chain
	Striking obstruction	Replace chain; eliminate interference(s)
	Cotters installed improperly	Correct installation

Source: IADC Drilling Manual, chap. G, sec. 3, p. 1.

To summarize—

Inspect the following on the selective transmission (and the compound on mechanical rigs) every tour:

- Lubrication system
- Damaged chains and sprockets
- Worn chains and sprockets
- Misaligned shafts or sprockets
- Chain that is too loose or too tight
- Damaged chain guards

Clutches

Clutches are the components of a mechanical assembly that connect or disconnect driving shafts from driven shafts. When a clutch is engaged, it makes the connection so that the driving shaft moves the driven shaft. When a clutch is disengaged, it breaks such a connection. Then, even though the driver continues to move, the driven shaft stops moving. A clutch works like an on/off switch for the transmission.

Locations

The hoisting system has clutches wherever two drives are connected and the driller needs to be able to disconnect them (see fig. 18). An exception to this is when a hydraulic coupling or a torque converter replaces the mechanical clutch. Some of the places that need clutches are—

1. between the compound and the mud pumps;
2. between the compound and the selective transmission (master clutch);
3. inside the selective transmission, for example, to the high and low drum drives (high clutch and low clutch), to the sand reel;
4. between the selective transmission and the rotary drive.

The type of clutch used in each location depends on its position in the machinery, the space available, the conditions under which it must function, the possibility for misalignment, and the work that must be done. Three types of clutches in the drawworks are the positive clutch, the friction clutch, and the overrunning clutch.

Positive Clutches

A clutch always has two parts, one for each of the two shafts it is connecting. A *positive clutch* is the simplest type of clutch. Spline and jaw clutches are two types of positive clutches in the drawworks. A *jaw clutch* has two jaws, one on the end of each shaft, that interlock when pushed together (fig. 41a). A *spline clutch* is a shaft with grooves in it and a cylinder with matching grooves or teeth that fits over the shaft (fig. 41b).

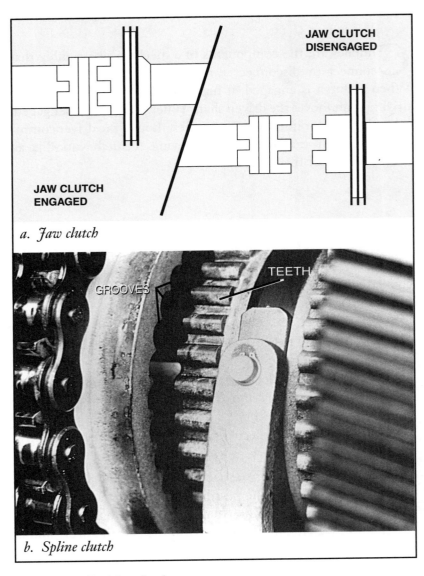

a. Jaw clutch

b. Spline clutch

Figure 41. Positive clutches

Positive clutches do not slip; they are either engaged or disengaged. The jaws or splines must align perfectly in order to mesh, so the driller cannot engage a positive clutch until both the driving and the driven shafts are stopped or moving only very slowly. The shock of engaging the clutch while the shafts are moving may break splines and teeth or twist the shaft. The shock might also cause the chain of a chain-and-sprocket drive to break.

Positive clutches work best in components of the hoisting system where the clutch does not require frequent engagement and disengagement and is in the presence of oil lubrication. Lubrication helps the two halves move into place smoothly to lock together. The selective transmission uses spline clutches in pressure-lubricated chain-and-sprocket drives because the oil spray or mist hits the clutch as well as the chain.

Sometimes one of the clutches between the drawworks drumshaft and a hydrodynamic auxiliary brake is a jaw clutch. The compound also may use a jaw clutch as a connection to the mud pumps.

Friction Clutches

Instead of having two toothed sections that mesh, a *friction clutch* has two smooth surfaces, either flat plates or a drum and sleeve, lined with special material that produces a lot of friction. Imagine slowly putting a phonograph record down onto a rotating turntable. As you lower the record, it starts to catch and turn along with the turntable. The closer the record gets to the turntable, the more friction there is between them. The record gradually turns faster until it is turning the same speed as the turntable. In the same way, when the driller engages the clutch, the clutch surfaces squeeze together. The end attached to the spinning (driving) shaft gradually catches the end attached to the stationary (driven) shaft. The stationary shaft starts to rotate because of friction until its speed is the same as the engine speed. This is a case of useful, rather than harmful, friction.

Unlike in a positive clutch, the two halves of a friction clutch can slip and engage gradually, so there is no shock to either shaft at any speed. Oil on a friction clutch produces too much slipping, so the clutch never engages completely, and the driven shaft never reaches the speed of the driving shaft. For this reason, friction clutches cannot work in the presence of oil. So any two drives that are completely enclosed in a chain guard or housing with pressure or disk lubrication must use a jaw clutch.

Drum clutches and plate clutches are types of friction clutches that a drawworks uses.

Drum Clutch

A drum clutch has a metal housing lined with an expandable diaphragm, like an inner tube, and pads of friction material called friction shoes (fig. 42a). This whole assembly is bolted to the driven shaft. An air compressor on the rig floor with controls at the driller's console forces air into the diaphragm, which expands like a balloon. This works like filling up a tire at the gas station with the air hose. The pressure of the expanded diaphragm squeezes the friction shoes against a drum attached to the driving shaft. Friction between the shoes and the drum allows the clutch to engage. The clutch disengages when the driller lets the air out of the diaphragm.

The high drum drive often uses a drum clutch. The drives to the mud pumps from the compound, the sand reel drive, and the rotary countershaft drive may also use drum clutches.

A plate, or disk, clutch has metal plates that may or may not have a friction material on them (fig. 42b). Like the drum clutch, it also has an air-operated diaphragm that presses the plates together. Springs push the plates apart when the clutch is disengaged. The low drum drive may use a plate clutch.

Plate (Disk) Clutch

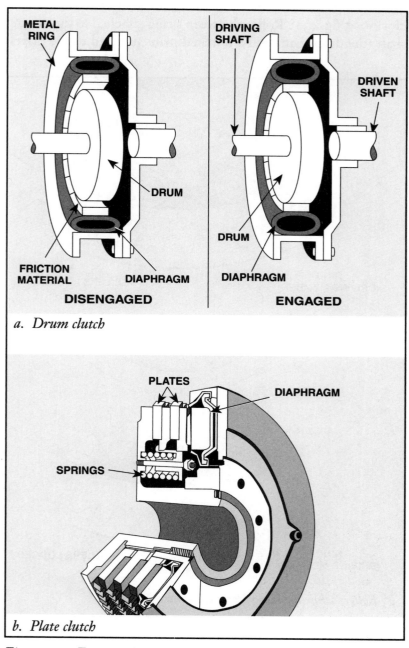

a. Drum clutch

b. Plate clutch

Figure 42. Friction clutches

Overrunning Clutches

An overrunning clutch is a special type of clutch that engages and disengages automatically. Thus, the driller has no control over when it is working. It connects a hydrodynamic auxiliary brake (see "Auxiliary Brakes") to the drum so that the brake engages whenever the driller is lowering the traveling block. Look at figure 43 to understand how an overrunning clutch works. When the drum is hoisting the traveling block, it turns in one direction (counterclockwise, fig. 43a). Rollers between a ring attached to the driving shaft (the drumshaft) and a notched plate attached to the driven

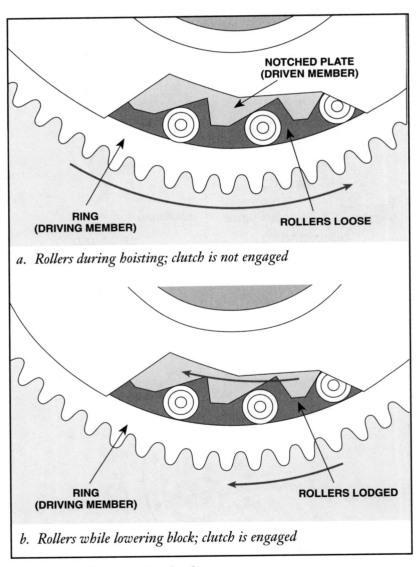

a. Rollers during hoisting; clutch is not engaged

b. Rollers while lowering block; clutch is engaged

Figure 43.   Overrunning clutch

shaft (the brake shaft) roll the same direction as the drum, counter-clockwise. They slip along the slope of the irregularly shaped notches on the driven plate, and the brake shaft does not move. When the drum turns in the other direction to lower the block (clockwise, fig. 43b), however, the rollers catch in the other side of the notches on the driven plate. Now they cannot roll, and the brake shaft turns clockwise with the drum. This activates the brake.

The hydrodynamic brake may have a jaw clutch as well, so that the driller can completely disconnect the brake during drilling.

After installing a new friction clutch, adjust the distance between the friction surfaces (plates or drum and shoes). They need to be close enough to engage when the diaphragm expands, but not so close that they touch and drag when the diaphragm contracts. The manufacturer provides instructions for doing this. A new disk clutch requires several adjustments in the first few hours of use until the plates wear in and mate properly.

Installation

Maintenance

Clutches wear because of friction between the moving parts and because of misalignment. Regular maintenance can assure a long life.

Alignment

Chains and sprockets are not the only, or even the most important, parts that wear due to misaligned shafts. The presence of a friction clutch is an even more important reason to keep the shafts aligned.

Imagine a turntable that has no center pin to fit into the record's hole to center the record. If you set the record down off-center, it will move back and forth instead of just spinning smoothly. Likewise, if the two shafts going into a friction clutch are not aligned, the faces of the clutch that they fit into do not match up. This is *offset misalignment*. Even a small misalignment causes a vibration at the high rotation speeds in a drawworks, which stresses the clutch and the shaft and whatever is on the other end of the shaft, such as a sprocket and its chain or a gear.

Angular misalignment is like keeping the record centered on the turntable but tilting the top of the record to one side. The record starts wobbling back and forth because it is no longer flat against the turntable. When the two halves of the clutch, like the record and turntable, are not flat against each other, they wear out prematurely, and again, vibration stresses the shaft at both ends.

Lubrication and Cleaning

Friction and overrunning clutches have bearings where the shaft and the clutch meet. As with all bearings, lubricate them as often as the manufacturer recommends. However, because a friction clutch cannot work if there is oil on the friction surfaces, inspect the oil seal that holds the lubricant in each bearing assembly. If it is no longer functioning, the clutch will slip and fail to transfer the power (the torque of the driving shaft) effectively. Replace the seal if it is worn out.

Sometimes the clutch's friction surfaces get dirty, which causes the clutch to hang up and drag when it is disengaged. The dirt may be dust from the environment or from the friction material itself wearing. Take the clutch apart to clean it, following the manufacturer's instructions.

To summarize—

Types of clutches in a drawworks
- Positive—jaw or spline
- Friction—drum or plate
- Overrunning

Where the types work
- Positive—in the presence of oil; between drives that do not have to be disconnected often
- Friction—away from oil; between drives that need to be engaged and disengaged frequently
- Overrunning—where automatic engagement and disengagement are desirable, usually to the hydrodynamic brake

Installing friction clutches
- Adjust the clearance between friction surfaces

Maintaining clutches
- Keep the shafts aligned
- Lubricate bearings
- Inspect oil seals
- Keep the moving parts clean

Main Brake

The main brake is crucially important to a drilling rig because it slows or stops the drum. It is also called a mechanical brake, because it uses only mechanical energy rather than electricity or water power to work. The crew must adjust it, service it, and reline it regularly and should therefore be thoroughly familiar with its construction and operation.

Design

Figure 44 shows the main brake. The figure does not include the drum and its rims, which are also a part of the brake, so that the other parts are easier to see. The bands wrap around the rims of the drum and have a lining of brake blocks to increase friction. The driller applies the brake by pulling down on the brake lever, which is next to the console.

Brake Flanges and Bands

The drawworks drum is just a cylinder to start with. The manufacturer bolts steel rims to the ends of the cylinder that make it into a spool, and these rims are half of the mechanical brake (see fig. 6). The rims are also called *flanges*. These flanges have a hardened layer on the surface that does not wear out quickly.

The other half of the brake is the *brake bands*. These are two flexible steel bands that wrap around the flanges. One end of each band (the *dead end*) is anchored to the drawworks frame and does not move. The other end (the *live end*) is attached to a brake lever by means of a linkage. Moving the brake lever pulls the live end down, and the whole band tightens around the flange. This slows or stops the drum by friction. The main brake, like the friction clutch, takes advantage of useful friction, where the energy of the moving drum is transferred to the immovable brake bands. This mechanical energy from motion has to go somewhere; here it changes into heat. Remember that converting energy from one form to another is one of the things machines can do.

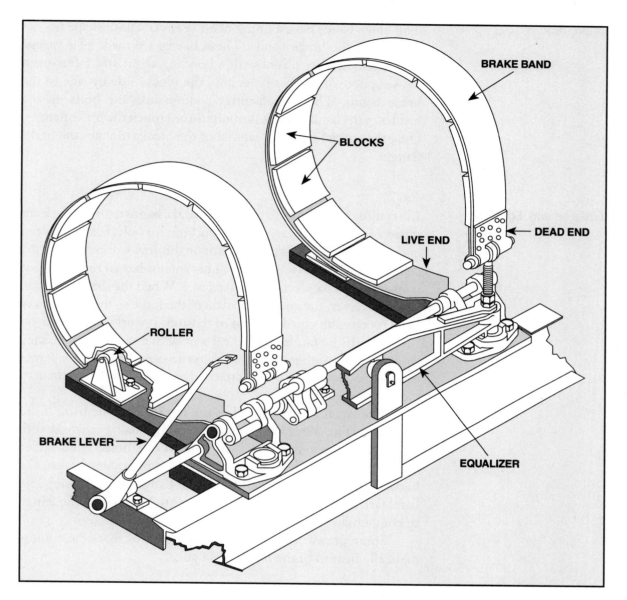

Figure 44. Main (mechanical) brake

The brake bands are heavy and eventually they sag and get out of round. For this reason, each band has one or two rollers beneath it to push the bottom up toward the rim and often springs above it (not shown) that pull up the top.

Brake Blocks

The brake bands have a lining of *brake blocks* whose shape fits the contour of the brake bands. These blocks are made of a special heat-resistant fiber mixed with a bonding agent and interwoven with copper wire. Brass bolts hold the blocks side by side to the brake bands. The manufacturer countersinks the bolts on the insides of the bands so that the bolts do not touch the brake flanges. Only the special friction material of the blocks touches the brake flanges.

Linkage and Equalizer

Everything from the dead ends of the brake bands to the brake lever is part of the *brake linkage*. Heavy-duty pins or bolts fasten the dead ends of the brake bands to an anchor on the drawworks frame at the center of the drum (see fig. 44). This anchor has an *equalizer* that connects both bands to one brake lever. When the driller pulls on the brake lever, the equalizer balances the force so that each brake band receives an equal amount of tension. It works like a yoke on two oxen. If the farmer pulled each ox with a separate rope, they would not move together. But when the oxen have a yoke across their necks, pulling on a rope attached to either side forces them to move at the same time.

The equalizer also switches the full load to one band if the other fails. The crew must adjust this type of equalizer to make sure that it does not assign all of the load to one brake band under ordinary conditions. The adjustment consists of varying the distance between the brake band and the brake flange on each side. The band farthest from the brake lever needs to be closer to the flange to compensate for being farther from the force that moves it.

Some drawworks have equalizing devices that adjust automatically instead of the mechanical yoke.

How does the manufacturer decide which side of the drum to attach the dead end of the brake band? The answer has to do with the fact that the brake works better in one direction than the other. Because gravity is always pulling the traveling block and the drill stem down, the mechanical brake needs to work best in that direction.

The braking action is stronger toward the dead end than at the live end. When the driller pulls the brake lever, the part of each brake band nearest the dead end comes into contact with the brake flange first (fig. 45). Because of this, the brake functions better when the drum is rotating in one direction than in the other. In figure 45, counterclockwise rotation sends the drum upward into the strongest braking action (fig. 45a). Clockwise rotation is, in a sense, downward relative to the area of strongest braking, and the brake will not stop as efficiently in this direction (fig. 45b).

Manufacturers design the drawworks so that when the brake lever is on the front side of the drum, the drum rotates in the correct direction for the best braking as the block descends.

When the drum is hoisting the block, the brakes take longer to stop it. If the driller pulls too hard on the brake lever to try to stop the drum quickly when hoisting, the drilling line on the drum can backlash and unseat the clamp that attaches the line to the drum.

Operation

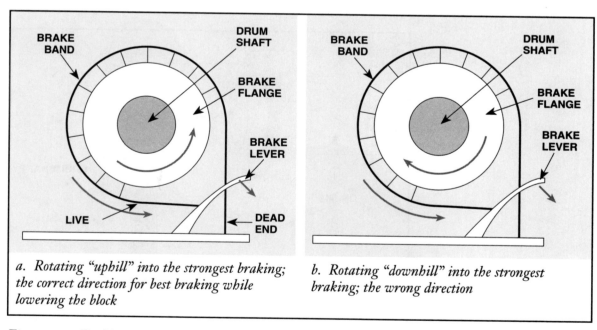

a. Rotating "uphill" into the strongest braking; the correct direction for best braking while lowering the block

b. Rotating "downhill" into the strongest braking; the wrong direction

Figure 45. Braking capacity increases from the live end to the dead end.

Braking Capacity

Several factors affect braking capacity (that is, how much weight the brake can stop in a given length of time). All have to do with the size of the braking area. A larger area means better braking.

The first factor is the ratio between the diameter of the brake rim (the brake flange) and the diameter of the drum (fig. 46). If two drums are the same size, but the flanges of one are larger in diameter and therefore have a longer circumference, the one with the larger flanges will brake better because it has a larger braking area. In effect, you are using a larger brake for the same-sized drum.

Another factor in braking capacity is the width of the brake band (fig. 47). A narrow band will hold as much weight as a wide one initially, but it will heat up faster. This happens because the narrow band has a smaller area with which to absorb the mechanical energy. When the brake blocks get hot, the fiber they are made of gets slicker, producing less friction and therefore less braking capacity.

A third factor in braking capacity is the *angle of wrap*. The angle of wrap is the distance in degrees that the brake band wraps around the flange (fig. 48). A full angle of wrap is 360°, all the way around the rim. Drawworks always have an angle of wrap of at least 270°, or three fourths of the way around. Other factors being equal, the greater the angle of wrap, the greater is the braking capacity, again because more of the band contacts the flange.

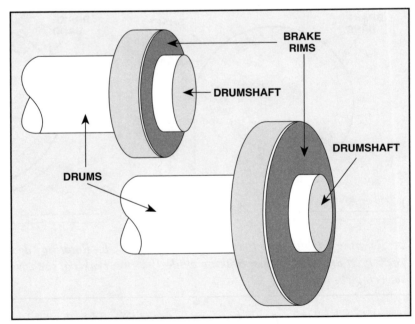

Figure 46. Comparing the diameters of the drum and the rims

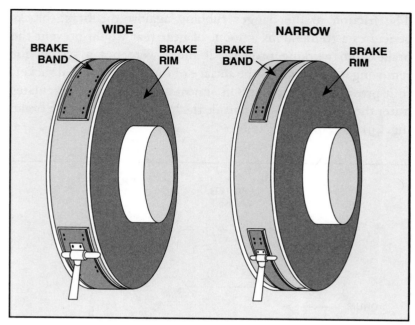

Figure 47. Comparing the width of the brake band

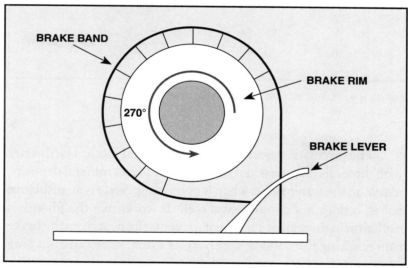

Figure 48. Angle of wrap of 270°

Cooling   The friction of the flanges rubbing against the brake blocks generates a tremendous amount of heat that would prevent the brake from working properly if there were not a system for removing it. Heat makes the surface of the lining material slicker, so it provides less friction. On almost all rigs, a pump circulates water through a *water jacket* inside the brake rims to cool the brake (fig. 49).

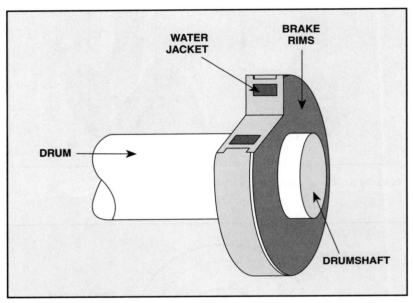

Figure 49.   Cross section of the brake rim

The circulating water must be clean and fresh. Hard water (with minerals dissolved in it) or salt water leaves mineral deposits, or *scale*, in the water jacket when it evaporates. Scale is an insulator; that is, it does not conduct heat well. It works like the fiberglass insulation in the wall of a house: it prevents the heat from the brake from reaching the cooling water. At its worst, scale can even plug up the water jacket entirely so that no water flows through it. Salt water may cause corrosion, which damages the metal parts. Water leaking out of any part of the cooling system is a sign that the part may be corroded. Both scale and corrosion can ruin the water circulation system and prevent the brake from working.

The power for the cooling water pump cannot come directly through the drawworks because while the brakes are working the hardest and have the greatest need for cooling water—during tripping in—the master clutch is usually disengaged. On large rigs, the prime movers directly power the water cooling pump. Some older or smaller rigs may use an independent motor or diesel engine.

The lines through which the water flows to and from the brakes are large because friction between the water and the line can slow down the flow. The outlet line should be lower than the brake, also to keep the water flowing freely downhill out of the brake.

It is crucial that the driller turn on the brake's cooling water supply *before* using the brake. If the driller turns on the water supply after the brake has already heated up, the rims and water jacket can crack or break, just like a hot light bulb plunged into cold water will shatter.

Safety Shutdown Device

The driller must watch the traveling block's movement when hoisting and keep it from hitting the crown block on its trip upward. But as a safety measure, many rigs also have an air-operated, or pneumatic, shutdown (commonly called a Crown-O-Matic™) to prevent this from happening. The shutdown includes a sensor for determining when a certain amount of drilling line has spooled onto the drum. At that point, the sensor triggers a lever on the drum that automatically shuts off the power to the drum and locks the brake down. The driller must reset the lever and bleed off the air pressure before hoisting the traveling block again.

To summarize—

Components of the main brake

- Flanges (rims)—the rims of the drum
- Bands—flexible steel bands that wrap around the flanges
- Blocks (lining)—blocks of a friction material that line the bands
- Linkage—the mechanical connections between the bands and the brake lever
- Equalizer—a mechanism that assures that both bands brake the drum equally when the driller pulls the brake lever

Factors in braking capacity

- Diameter of the rims
- Width of the bands
- Angle of wrap

Components of the cooling system

- Water jacket—a passageway inside the rim for circulating water
- Clean, fresh water
- Water pump—powered by an independent motor or engine
- Inlet and outlet lines

Manufacturers of drawworks usually furnish complete instructions for maintaining the mechanical brake. The owner or operator of the rig should be familiar with the complete and specific details of maintenance and safety considerations. But the crew still needs to understand the basics. Fortunately, a trained floorhand can easily recognize most signals that indicate the need for maintenance. Since adjustments and replacement of brake blocks are routine with rig crews, the precautions and planning involved are largely just normal care and common sense.

As a practical matter, the overriding concern in brake system maintenance is monitoring of the various parts to assure that wear or deterioration does not progress beyond safe limits.

Maintenance

Brakes wear out because of the friction between the various parts. The flanges rub against the blocks, or lining, and both wear down. The crew should perform the following maintenance procedures on the main brake:

1. lubricate and visually inspect the parts on a set schedule;
2. adjust the brake handle position;
3. adjust the band rollers and springs;
4. inspect the lining (blocks) and brake rims for wear;
5. inspect the cooling system;
6. inspect the bands; especially after replacing the blocks;
7. inspect the linkage (live and dead ends) for wear.

The experience of the operator, rig conditions and usage, and recommendations of the manufacturer will influence how often the crew makes these inspections and adjustments. Most of them are part of daily rig operation. Any unusual performance of the brake requires special or more frequent inspections. Keep records of all inspections and repairs.

General Maintenance Procedures

Brake Bands   If a brake band fails, the hoist has no main brake, a potentially catastrophic situation. Thus, proper maintenance and regular inspections of the bands are critical to safety. Using a damaged brake band can also cause the brake lever to kick when the driller releases it, which can injure the driller. A damaged band can also cause the linkage components to break.

The largest contributor to wear on brake rims and lining, other than normal use, is brake bands that are not round. The bands can bend, twist, kink, or flatten. The cause may be an improperly adjusted equalizer or rough handling of the band when installing or removing it. Gravity also pulls on the bands and causes them to get out of round. Inspect the rollers and the springs to see that they are all in place and working properly. The rollers must roll, and the springs must still be springy. The rollers and the springs should maintain a clearance between the blocks and the rims when the brake is not engaged.

Always take off the brake bands by hand. Do not use the catline or any other similar means for this. Jerking the brake band with the catline can cause kinking and distortion. Store a new brake band flat on a level surface to keep gravity from distorting it.

One way to check a brake band for roundness is by drawing a circle on the floor, using a nail and string, for example (fig. 50), and then laying the band on it. If it is slightly out of round or has a small kink, it is possible to straighten it by hammering on it. Hammering on metal stretches it, so stretching the band in the right places can repair a minor problem. If the band has a large kink or twist, replace it.

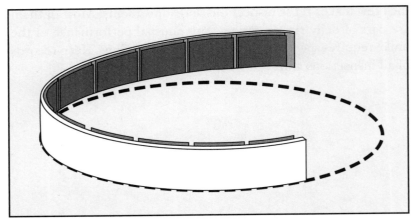

Figure 50.   Checking the band for roundness

Each time they remove the brake bands, crew members should inspect brake bands for the following problems as well as for roundness:

1. Cracks around holes, rivets, welds, and any other area of stress. These are called fatigue cracks and can cause the band to break. Before the crew relines a used band, a technician should always inspect it for invisible cracks by a method such as magnetic particle inspection or dye penetrant inspection. A band with fatigue cracks is unusable and not repairable. Never reuse it.

2. Nicks, gouges, dents, or other surface irregularities. Such defects act as *stress risers*, which means that they raise the level of stress at that point and cause fatigue cracks. A machine shop can file and smooth them off.

3. Welded repairs. Never repair a brake band by welding it. In particular, welds to connect two ends that run across the width of the band act as stress risers.

Brake Flanges

If the crew maintains them properly, brake flanges (rims) should last for a year or more of active service. Each time crew members replace the brake blocks, they should also inspect the brake flanges. The best time to schedule maintenance on the brake rims and the lining is when the rig is idle.

Inspecting the brake rims first involves measuring their thickness. Manufacturers design brake rims to wear down a prescribed amount before they are no longer functional.

As brake flanges wear, debris such as sand or salt between the lining and the flanges causes scoring or grooving in the flanges. The crew should measure the wear at the deepest grooves to determine whether the bands are still safe to use. The simplest way to do this is to place a straightedge or carpenter's square across the top of the rim and measure the depth of the grooves with a depth gauge (fig. 51). The straightedge rests on the lip of the rim. Subtract the height of the lip above the rim when new from the depth measurement to get the amount of wear. The maximum allowable wear can range from ¼ inch to ⅞ inch (6 to 22 millimetres), depending on the manufacturer. If the amount of wear is greater than the manufacturer's recommendation, send the drum to a machine shop to repair or replace the flanges.

The flanges wear out quickly once the hardened layer on their surface has worn away. The rig operator may hire an inspection company to conduct ultrasound or hardness tests to discover this.

Also check rims for scale in the water jacket. Descale them with a chemical treatment if inspection reveals significant scale buildup.

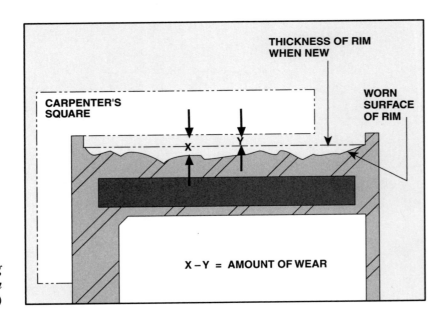

Figure 51. Measuring scoring in the brake rim (cross section)

Brake Lining

Under ordinary working conditions, a good-quality, properly installed brake lining (blocks) should serve through drilling operations on five or six medium-depth wells. The crew will need to replace the brake blocks when they have worn down to a certain thickness or when they have been on the brake a certain length of time. The manufacturer recommends the time and thickness limits. If the wear is uneven, determine why and repair that problem before returning the drawworks to service and ruining a new lining.

The lining bolts will loosen as a new lining seats itself, so check them after installing a new lining and retighten them as required. Flattening the ends of the bolts (called peening) will keep the nuts from coming off but will not guarantee that the bolts will stay tight, so check them regularly.

Sometimes a supplier will deliver replacement linings already installed on a set of used brake bands. In this case, inspect the used bands carefully for damage.

Protect new blocks by storing them in a clean, dry place away from high heat and sunlight, which can cause the lining material to deteriorate.

Brake Linkage

The brake linkage is a system of mechanical parts that connect the bands to the brake lever. Friction wears the moving parts of the brake linkage, as it does any other metal part, so it needs lubrication. As it wears, the mechanism gets slack. This affects the adjustment of the linkage (see the next section) and therefore the operation of the brakes. Several factors can accelerate this wear, including infrequent lubrication, use of improper greases, grit in the grease, and extreme operating temperatures.

The crew must regularly inspect the linkage for fatigue cracks as well as for wear. The drawworks manufacturer provides information on the safe limits of wear.

After restoring the flanges, the lining, and the bands to good working order, reassemble the brake and make the following adjustments. Begin with low tension on the brake bands. Before giving the traveling block a full load, raise and lower it empty a few times while applying pressure to the brake lever in order to fit the new linings to the new flanges.

Then check and adjust the clearance between the bands and the flanges with the brake released.

As the band wears, it stretches. Gradually, the driller must pull the brake lever farther and farther down before the brake works. Adjust the linkage as needed. Adjust rollers and springs according to the manufacturer's specifications.

To summarize—

Maintaining the main brake
- Inspect the bands for roundness, cracks, gouges, and welded repairs
- Check the flanges for wear and the water jacket for scale
- Inspect the blocks for wear
- Adjust the brake lever
- Adjust the rollers and the springs
- Inspect the linkage for wear
- Grease the linkage on a regular schedule

Auxiliary Brake

▼
▼
▼

The auxiliary brake works in combination with the main brake to slow the rate of descent of the traveling block with a heavy load. It functions only when the block is descending. The auxiliary brake ensures that the load descends slowly and smoothly and it lessens wear on the main brake by taking the heavy shock loads (sudden jerking) and continual dead weight off the brake bands.

Always be careful, however, to lower the traveling block slowly enough that the mechanical brake alone could stop it, because the auxiliary brake could fail. The auxiliary brake cannot stop the drum by itself.

The auxiliary brake can be either a hydrodynamic brake (activated by water) or an electrodynamic brake (activated by electricity). It sits next to the drum on the drawworks framework (see fig. 5a). The two types look very similar on the outside.

Hydrodynamic Brake

Rig workers often call a hydrodynamic brake by a well-known trade name, *Hydromatic®*. It works on the same principle as the hydraulic coupling between the compound and the prime mover. The hydrodynamic brake consists of a rotor on a shaft inside a housing full of water (fig. 52). The water in the brake resists the movement of the rotor, which turns because its shaft is connected to the drumshaft. This slows down the drumshaft and therefore the drum.

The driller can vary the braking capacity by changing the water level. The brake has either a valve that varies the amount of water pumped in or holes with stoppers that can be removed to let water out.

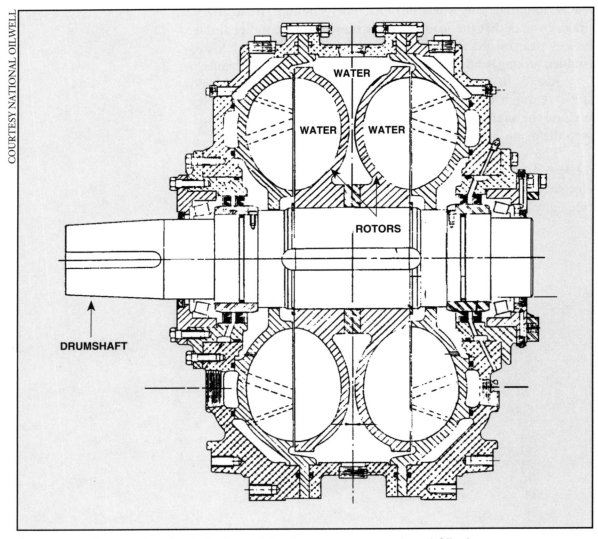

Figure 52. Cross section of a hydrodynamic brake. Water surrounds and fills the rotors.

Between the drumshaft and the rotor is an overrunning clutch. This clutch allows the hydrodynamic brake's rotor to turn with the drum every time the traveling block is going down. When the driller hoists the traveling block, the clutch disengages and freewheels. The brake rotor remains stationary and does not hinder the hoisting operation.

The energy from the moving drum converts to heat in a hydrodynamic brake, like in the main brake, so the water circulating through the brake gets hot. The amount of water and how fast it flows through the brake determine how much heat builds up. Usually the hydrodynamic brake uses the same water and pump as the main brake. The water must be fresh (not hard or salt) and free of foreign substances, such as sand.

Water Circulation and Cooling

Using a hydrodynamic brake makes maintaining the circulation system even more important, since water is what makes the brake work. In addition to inspecting the water lines, the crew must inspect the brake itself for corrosion and scale. Check the water level regularly as well. Heat can cause the water to evaporate or boil off. The temperature at the brake outlet must never exceed 180°F (82°C).

Maintenance

Electrodynamic Brake

An electrodynamic brake (also called an electromagnetic brake) uses a different type of resistance to slow the drum, but it is not so different from a hydrodynamic brake in principle. Rig workers often call an electrodynamic brake by a well-known trade name, *Elmagco®*. Instead of the drumshaft's connecting to a rotor shaft, it connects to the shaft of a cylinder with wire coiled around it (an armature), which thus rotates with the hoisting drum (fig. 53). The coiled wire of the armature is an electromagnet, which means that it becomes a magnet when electricity is flowing through it. When the electricity is shut off, it is just a plain wire. In a housing surrounding the armature is a set of electromagnets that do not move.

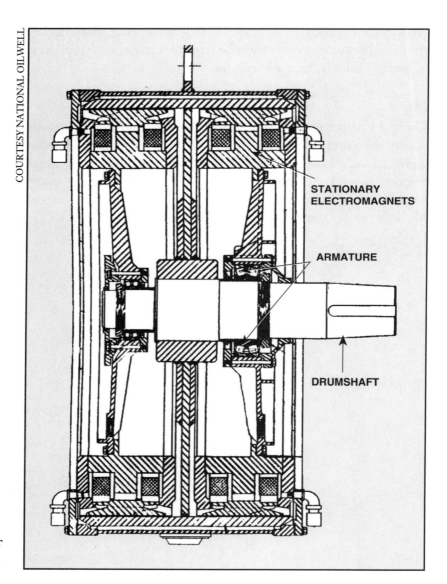

COURTESY NATIONAL OILWELL

STATIONARY ELECTROMAGNETS

ARMATURE

DRUMSHAFT

Figure 53. Cross section of an electrodynamic brake

When the current is on, the armature's electromagnet acts like the north pole of a regular magnet, and the stationary electromagnets act like the south pole. They resist each other, just as the north and south poles of ordinary bar magnets do. This resistance slows down the armature and therefore the drum. The strength of the magnetic fields for both electromagnets varies according to the amount of electric current passing through them. The driller controls the amount of braking action by adjusting the strength of the magnetic fields, like turning a dimmer switch.

The auxiliary brake does not need to back up the main brake while the rig is drilling. At this time, the driller can disengage the electrodynamic brake by means of a coupling to protect it from the vibrations of drilling. The shaft connection does not need a clutch to disconnect this brake during hoisting because, when the electric current is shut off, the armature does not brake; it just rotates freely with the drum.

Cooling

An electrodynamic brake also has a cooling system of water circulating through it to remove heat from transferred mechanical energy. As in the hydrodynamic brake, it is part of the main brake's circulation system.

Maintenance

Maintenance of an electrodynamic brake is minimal. Except for the roller bearings that support the shaft of the armature, there is no physical contact between the armature and the stationary housing. These bearings are oversized and last for years of continuous operation. With no other surfaces to wear, the performance of an electrodynamic brake does not deteriorate, even with long, hard use. However, the crew must regularly check the circulating system for corrosion and scale and check the cooling water level and quality. Follow the manufacturer's suggestions for lubricating the bearings.

To summarize—

Types of auxiliary brakes

- Hydrodynamic (Hydromatic™)—uses water to brake; works automatically when lowering the block
- Electrodynamic—uses electromagnets; works by switching on electric current

Maintaining the auxiliary brake

- Check the water level in the circulation system
- Inspect the circulation system for scale and corrosion
- Grease the bearings in an electrodynamic brake

Catshaft

The *catshaft* is a long axle that sits on heavy-duty roller bearings in the frame of the drawworks. It runs along the top of the drawworks and sticks out on both sides of it (fig. 54). The catshaft has two catheads on each end and often a sand reel in the middle. Each cathead is a winch that can spool up a wire rope, fiber rope, or chain. The catheads are an integral part of the catshaft and rotate with it.

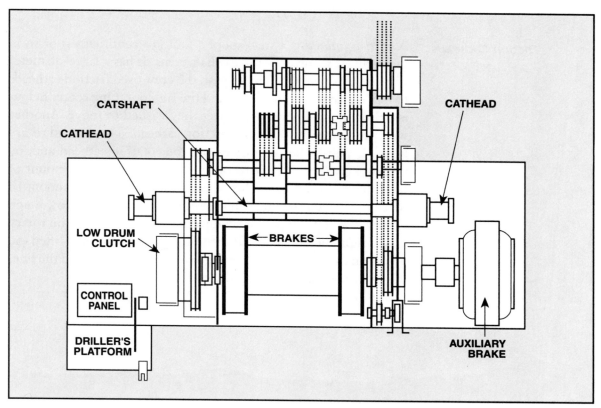

Figure 54. The catshaft, catheads, and sand (coring) reel

The power for turning the catshaft comes from the drumshaft by means of a chain-and-sprocket drive. The catshaft has a clutch with safety locks that keep the clutch disengaged until the driller specifically overrides them. This prevents the driller from accidentally engaging the catheads and injuring workers.

Catheads

The four catheads are of two types: friction catheads and automatic, or mechanical, catheads. The two automatic catheads help the driller make up and break out drill pipe. In the past, the two friction catheads helped move heavy pieces of equipment around the rig floor. Now, however, rotary helpers use small air-activated hoists for this purpose. The catshaft has one automatic cathead and one friction cathead on each end.

Friction Catheads

A *friction cathead* is a steel spool a foot (30 centimetres) or so in diameter. It revolves as the catshaft turns. It has a large-diameter fiber rope as its *catline*. In the past, the crew used friction catheads to move heavy equipment around the rig floor. One rotary helper attached the catline to the object they wished to move. Another wrapped the line around the friction cathead. This second rotary helper grabbed the catline nearing the rotating cathead and, by pulling hard or not so hard on the line, adjusted the amount of tension on the catline. This tension determined the amount of friction between the line and the rotating cathead and how much or little the line slipped or held to the cathead. When the rotary helper tightened the line, the cathead pulled the line and lifted the object on the other end. When the rotary helper released the line, the cathead stopped pulling the catline.

Today, even though most drawworks still have friction catheads on them, most contractors substitute air hoists, or *tuggers*, which are on the rig floor separate from the drawworks (fig. 55). Tuggers are safer and easier to use than friction catheads. On a tugger, a small, powerful pneumatic motor powers a spool that takes in or lets out wire rope. The wire rope is about ½ inch (13 millimetres) in diameter, and it lasts longer than fiber rope. After one rotary helper rigs the line to an object, another starts the tugger motor to pull the line and move the object.

Figure 55. Air tugger

Automatic Catheads

The two automatic catheads are reels on either end of the catshaft, inward from the friction catheads. The makeup cathead is on the driller's end of the catshaft; the breakout cathead is on the other end.

The catline on automatic catheads is usually called a tong pull line. A tong pull line is either a wire rope or a chain. The driller controls the automatic catheads from the console with a lever. The lever connects to built-in friction clutches on the cathead drums. The clutches give the driller precise control over the force of the line pull. Automatic catheads do not need brakes; they are either on or off, depending on whether the clutch is engaged or not.

The *makeup cathead* helps to screw together joints of pipe or drill string connections when pipe is being run into the hole. One end of a Y-shaped chain, like an ordinary chain only stronger, is attached to the cathead. Of the two remaining ends, one is longer than the other. A rotary helper attaches the longer end to the makeup tongs and then latches the tongs around the tool joint. The third, shorter end of the chain wraps around the tool joint. When the driller engages the makeup cathead, it spools in the chain, and the end of the Y wrapped around the joint spins the pipe until the chain completely unwinds. Then the cathead takes up the slack in the longer end of the Y and tightens the joint. It works something like spinning a nut with your fingers until it catches and then using a wrench to tighten it.

The *breakout cathead*, located opposite the driller's position, looks exactly like the makeup cathead except that its line is a wire rope. Rotary helpers use it to unscrew tool joints as they come out of the hole. Just as the makeup cathead pulls on the makeup tongs to tighten and make up pipe, the breakout cathead pulls on the tongs to loosen and break out pipe. Wire rope is stronger than chain of comparable size, will not stretch like a chain if overtorqued, and works better to break the screw joint loose.

Sand Reel

A *sand reel* is a relatively light-duty hoist for raising and lowering small tools such as a bailer or a logging device into and out of the hole (see fig. 56). The sand reel gets its name from the bailer, which removes sand, among other things, from the hole. The sand reel also can be used to retrieve cores, hence its alternate name, *coring reel*.

On drawworks that have a sand reel, the sand reel shaft sits on roller bearings on the catshaft. The spooling line, or *sandline*, on the sand reel is a wire rope about ⁹⁄₁₆ inch (15 millimetres) in diameter and several thousand feet or metres long. The sand reel's drum, like the main drum, has a mechanical brake, with brake bands on its flanges. Beneath one of the brake bands is a clutch that engages the sand reel with the catshaft. The driller controls it from the console.

To summarize—

Components of the catshaft assembly
- Catshaft
- Friction catheads
- Automatic catheads
- Sand (coring) reel

Job of each component
- Catshaft—an axle for the catheads and sand reel
- Friction catheads—spools used for moving heavy objects on the rig floor; now replaced by tuggers
- Automatic catheads—reels for making up and breaking out drill pipe
- Sand (coring) reel—a reel on some, but not all, catshafts for hoisting small tools or core samples out of the hole

Lubrication

▼
▼
▼

The greatest contribution the rig crew can make to the general benefit of the drilling contractor is to learn and practice good techniques of rig lubrication. Every place on the drawworks where metal rubs against metal needs lubrication. Good lubrication helps the equipment to last as long as possible before breaking. Not only does this mean less expense for replacement parts, but it also means that there is less time when the rig is not drilling because of repairs.

Every tour, the operator provides a certain amount of time for inspecting and servicing the rig. The driller keeps a maintenance record, called a tour report, that includes, for example, when the crew measured oil levels, changed oil, checked oil pressure gauges, serviced and replaced filters, and greased fittings.

Some parts of the drawworks, such as the transmission, need oil lubrication and some parts need grease. Figure 56 shows an

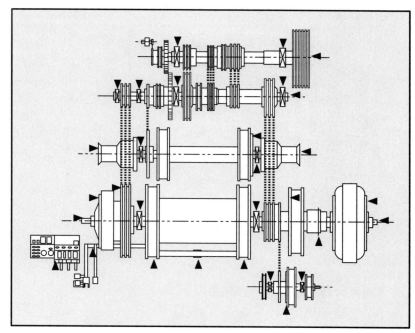

Figure 56. Grease fitting locations

example of the locations of grease fittings on one manufacturer's drawworks. The crew should grease all of these fittings every 24 hours, except the catheads. The catheads need grease before and after each trip. The type of grease depends on where it is used and on the weather temperature.

Any levers and linkages, like the brake linkage, need oil frequently so that they work easily and to prevent corrosion.

Drawworks manufacturers provide charts of recommended maintenance schedules, showing where to lubricate and how often, and what type of lubricants are best. Always follow these recommendations closely. Do not rely on old information, and certainly do not use information for a different type of equipment.

To summarize—
- Inspect the lubrication systems and grease fittings on a regular schedule
- Keep maintenance records
- Follow manufacturers' recommendations

Glossary

air hoist *n:* a hoist operated by compressed air; a pneumatic hoist. Air hoists are often mounted on the rig floor and are used to lift joints of pipe and other heavy objects.

American National Standards Institute (ANSI) *n:* serves as clearinghouse for nationally coordinated voluntary standards for fields ranging from information technology to building construction. Address: 11 W. 42d St., 13th floor; New York, NY 10036; (212) 642-4900.

angle of wrap *n:* the distance that the brake band wraps around the brake flange. Drawworks have an angle of wrap of 270° or more.

angular misalignment *n:* one type of misalignment in a chain-and-sprocket drive. The shafts are not parallel to each other (they form an angle) in either the horizontal or the vertical plane. This pulls the link plates on one side tighter than those on the other side; thus, one side of the chain and sprockets wears faster than the other. Link plates on only one side of the chain break because of fatigue.

ANSI *abbr:* American National Standards Institute.

automatic cathead *n:* see *breakout cathead, cathead, makeup cathead.*

auxiliary brake *n:* a braking mechanism on the drawworks, supplemental to the mechanical brake, that permits the lowering of heavy hook loads safely at retarded rates without incurring appreciable brake maintenance. There are two types of auxiliary brake—the hydrodynamic and the electrodynamic. In both types, work is converted into heat, which is dissipated through liquid cooling systems. See *electrodynamic brake, hydrodynamic brake.*

A

block *n:* any assembly of pulleys on a common framework; in mechanics, one or more pulleys, or sheaves, mounted to rotate on a common axis. The crown block is an assembly of sheaves mounted on beams at the top of the derrick or mast. The drilling line is reeved over the sheaves of the crown block alternately with the sheaves of the traveling block, which is raised and lowered in the derrick or mast by the drilling line. When elevators are attached to a hook on the traveling block and drill pipe is latched in the elevators, the pipe can be raised or lowered. See *crown block, traveling block.*

brake band *n:* a part of the brake mechanism consisting of a flexible steel band lined with a material that grips a drum when tightened. On a drilling rig, the brake band acts on the flanges of the drawworks drum to control the lowering of the traveling block and its load of drill pipe, casing, or tubing.

B

brake block *n:* a section of the lining of a brake band shaped to conform to the curvature of the band and attached to it with countersunk screws. See *brake band.*

brake flange *n:* the surface on a winch, drum, or reel to which the brake is applied to control the movement of the unit by means of friction.

brake linkage *n:* all equipment from the dead ends of the brake bands to the brake lever.

braking capacity *n:* how much weight the brake can stop in a given length of time

breakout cathead *n:* a device attached to the catshaft of the drawworks that is used as a power source for unscrewing drill pipe; usually located opposite the driller's side of the drawworks. See *cathead.*

bushing *n:* 1. a pipe fitting on which the external thread is larger than the internal thread to allow two pipes of different sizes to be connected. 2. a removable lining or sleeve inserted or screwed into an opening to limit its size, resist wear or corrosion, or serve as a guide.

C

casing *n:* 1. steel pipe placed in an oil or gas well to prevent the wall of the hole from caving in, to prevent movement of fluids from one formation to another, and to improve the efficiency of extracting petroleum if the well is productive. A joint of casing is available in three length ranges: a joint of range 1 casing is 16 to 25 feet (4.8 to 7.6 metres) long; a joint of range 2 casing is 25 to 34 feet (7.6 to 10.3 metres) long; and a joint of range 3 casing is 34 to 48 feet (10.3 to 14.6 metres) long. Diameters of casing manufactured to API specifications range from 4.5 to 20 inches (11.4 to 50.8 centimetres). Casing is also made of many types of steel alloy, which vary in strength, corrosion resistance, and so on. 2. large pipe in which a carrier pipeline is contained. Casing is used when a pipeline passes under railroad rights-of-way and some roads to shield the pipeline from the unusually high load stresses of a particular location. State and local regulations identify specific locations where casing is mandatory.

catenary *n:* the curve assumed by a perfectly flexible line hanging under its own weight between two fixed points. A suspension bridge is an example of a catenary structure; an anchor chain is a catenary.

cathead *n:* 1. a spool-shaped attachment on the end of the catshaft, around which rope for hoisting and moving heavy equipment on or near the rig floor is wound. 2. an automatic cathead. See *breakout cathead, makeup cathead.*

catline *n:* a hoisting or pulling line powered by the cathead and used to lift heavy equipment on the rig. See *cathead.*

catshaft *n:* an axle that crosses through the drawworks and contains a revolving spool called a cathead at either end. See *cathead.*

center distance *n:* in a chain-and-sprocket drive, the distance between the centers of the two sprockets or the shafts they fit on.

chain-and-sprocket drive *n:* a type of drive that consists of a loop of chain that goes around two shafts with sprockets fitted on the shafts. One shaft is the driving shaft (it provides the power) and the other is the driven shaft (it receives the power). When the sprocket on the driving shaft turns, the chain moves and turns the driven sprocket.

chain-and-sprocket transmission *n:* a selective transmission made up primarily of chain-and-sprocket drives. Between the input shaft coming from the compound and the output shaft on a mechanically driven drawworks are three two-strand chain-and-sprocket drives and a pair of gears for reversing. Each of these chain-and-sprocket drives has a different sprocket ratio to provide three speeds to the high drum drive and three to the low drum drive. Compare *gear transmission.*

chain case *n:* see *chain guard.*

chain guard *n:* in a chain-and-sprocket drive, a case that protects workers from the moving drive. It also protects the drive from dirt and is part of the cooling and lubrication systems. Some enclose only one drive and some enclose several drives. May be made of sheet metal or of plate metal. Heavier guards also support the shafts on bearings. Both types have access panels for inspection and maintenance.

chain width *n:* in roller chain, the width of the rollers, which is the distance between the inside faces of the roller link plates.

clutch *n:* a coupling used to connect and disconnect a driving and a driven part of a mechanism, especially a coupling that permits the former part to engage the latter gradually and without shock. In the oilfield, a clutch permits gradual engaging and disengaging of the equipment driven by a prime mover. *v:* to engage or disengage a clutch.

combined misalignment *n:* a type of chain misalignment that may result from combined angular and offset misalignment, or from two shafts that are not level with each other

compound *n:* 1. a mechanism used to transmit power from the engines to the pump, the drawworks, and other machinery on a drilling rig. It is composed of clutches, chains and sprockets, belts and pulleys, and a number of shafts, both driven and driving. 2. a substance formed by the chemical union of two or more elements in definite proportions; the smallest particle of a chemical compound is a molecule. *v:* to connect two or more power-producing devices, such as engines, to run driven equipment, such as the drawworks.

compounding transmission *n:* on a mechanical-drive rig, the type of transmission that sends power from the engines to the drawworks and the rotary table, and sometimes to the mud pumps. See also *transmission.*

connector link *n:* in roller chain, a type of link used to make a continuous loop of chain by connecting the two ends of the chain. The connector link is a pin link with either a spring clip or a cotter to hold the pins.

coring reel *n:* see *sand reel.*

cotter *n:* a spring steel wire that fits into an eye on a pin. It can be flared after insertion to hold the pin in place.

crown block *n:* an assembly of sheaves mounted on beams at the top of the derrick or mast and over which the drilling line is reeved. See *block.*

Crown-O-Matic *n:* a brand name for a special air-relay valve mounted near the crown that, when struck by the traveling block, conveys air pressure to the air brakes of the drawworks to prevent the traveling block from striking the crown.

D

dead end *n:* the end of a brake band that is anchored to the drawworks frame and does not move.

deadline *n:* the drilling line from the crown block sheave to the anchor, so called because it does not move. Compare *fastline.*

deadline anchor *n:* a device to which the deadline is attached, securely fastened to the mast or derrick substructure.

derrick *n:* a large load-bearing structure, usually of bolted construction. In drilling, the standard derrick has four legs standing at the corners of the substructure and reaching to the crown block. The substructure is an assembly of heavy beams used to elevate the derrick and provide space to install blowout preventers, casingheads, and so forth. Because the standard derrick must be assembled piece by piece, it has largely been replaced by the mast, which can be lowered and raised without disassembly. Compare *mast.*

diesel engine *n:* a high-compression, internal-combustion engine used extensively for powering drilling rigs. In a diesel engine, air is drawn into the cylinders and compressed to very high pressures; ignition occurs as fuel is injected into the compressed and heated air. Combustion takes place within the cylinder above the piston, and expansion of the combustion products imparts power to the piston.

disk lubrication *n:* see *slinger disk lubrication.*

drawworks *n:* the hoisting mechanism on a drilling rig. It is essentially a large winch that spools off or takes in the drilling line and thus raises or lowers the drill stem and bit.

drawworks brake *n:* the mechanical brake on the drawworks that can prevent the drawworks drum from moving.

drawworks drum *n:* the spool-shaped cylinder in the drawworks around which drilling line is wound, or spooled.

drawworks-drum socket *n:* a receptacle on the drawworks drum to which the drilling line is attached.

driller's console *n:* a metal cabinet on the rig floor containing the controls that the driller manipulates to operate various components of the drilling rig.

driller's side *n:* a panel, or console, on the left side of the drawworks (looking at it from the front).

drilling hook *n:* the large hook mounted on the bottom of the traveling block and from which the swivel is suspended. When drilling, the entire weight of the drill stem is suspended from the hook.

drilling line *n:* a wire rope used to support the drilling tools. Also called the rotary line.

drill stem *n:* all members in the assembly used for rotary drilling from the swivel to the bit, including the kelly, drill pipe and tool joints, drill collars, stabilizers, and various specialty items. Compare *drill string.*

drill string *n:* the column, or string, of drill pipe with attached tool joints that transmits fluid and rotational power from the kelly to the drill collars and bit. Often, especially in the oil patch, the term is loosely applied to both drill pipe and drill collars. Compare *drill stem.*

drip lubrication *n:* a method of lubricating a chain-and-sprocket drive in which a reservoir drips oil onto the chain at 4 to 20 drops per minute, depending on the speed of the drive. A drip lubricator for a multistrand drive has a pipe to distribute the oil to all strands.

driven shaft *n:* in a chain-and-sprocket drive, the shaft that receives the power. Compare *driving shaft.*

driving shaft *n:* in a chain-and-sprocket drive, the shaft that provides the power. Compare *driven shaft.*

drum clutch *n:* a type of friction clutch that has a metal housing lined with an expandable diaphragm and pads of friction material called friction shoes. The pressure of the expanded diaphragm squeezes the friction shoes against a drum attached to the driving shaft. Friction between the shoes and the drum allows the clutch to engage. See *clutch.* Compare *plate clutch.*

drumshaft *n:* in the drawworks, the axle on which the drawworks drum rotates. See *drawworks.*

E

electric rig *n:* a drilling rig on which the energy from the power source— usually several diesel engines—is changed to electricity by generators mounted on the engines. The electrical power is then distributed through electrical conductors to electric motors. The motors power the various rig components. Compare *mechanical rig.*

electrodynamic brake *n:* a device mounted on the end of the drawworks shaft of a drilling rig. The electrodynamic brake (sometimes called a magnetic brake) serves as an auxiliary to the mechanical brake when pipe is lowered into a well. The braking effect in an electrodynamic brake is achieved by means of the interaction of electric currents with magnets, with other currents, or with themselves.

energy *n:* the capability of a body for doing work. Potential energy is this capability due to the position or state of the body. Kinetic energy is the capability due to the motion of the body.

equalizer *n:* 1. a device used with mechanical drawworks brakes to ensure that, when the brakes are used, each brake band will receive an equal amount of tension and also that, in case one brake fails, the other will carry the load. A mechanical brake equalizer is a dead anchor attached to the drawworks frame in the form of a yoke attached to each brake band. Some drawworks are equipped with automatic equalizers. 2. any device used to distribute force equally on two pieces of equipment—for example, the pitmans on a beam pumping unit.

F

fastline *n:* the end of the drilling line that is affixed to the drum or reel of the drawworks, so called because it travels with greater velocity than any other portion of the line. Compare *deadline.*

fatigue *n:* the tendency of material such as a metal to break under repeated cyclic loading at a stress considerably less than the tensile strength shown in a static test.

flange *n:* a projecting rim or edge (as on pipe fittings and openings in pumps and vessels), usually drilled with holes to allow bolting to other flanged fittings.

forced lubrication *n:* see *pressure lubrication.*

friction *n:* resistance to movement created when two surfaces are in contact. When friction is present, movement between the surfaces produces heat.

friction cathead *n:* a type of cathead that helps move heavy pieces of equipment around the rig floor. See *cathead.* Compare *automatic cathead.*

friction clutch *n:* a clutch that makes connection by sliding friction.

friction shoe *n:* in a drum clutch, pads of friction material that are squeezed against a drum attached to the driving shaft. Friction between the shoes and the drum allows the clutch to engage.

G

gall *n:* damage to steel surfaces caused by friction and improper lubrication.

galling limit *n:* one of the limitations on chain-and-sprocket life. This limitation on the strength of the metal the chains and sprockets are made of may cause the metal to wear even though it is lubricated because of a heavy load or a high speed.

gear *n:* a toothed wheel made to mesh with another toothed wheel.

gear transmission *n:* in a small drawworks, the type of transmission used for lighter-duty hoisting because it takes up less room than chain-and-sprocket drives; generally have a torque converter instead of a clutch at the connection to the compound. Compare *chain-and-sprocket transmission.*

generator *n:* a machine that changes mechanical energy into electrical energy.

H

high drum drive *n:* the drive for the drawworks drum used when hoisting loads are light.

hoisting system *n:* drawworks, drilling line, and traveling and crown blocks. Auxiliary hoisting components include catheads, catshaft, and air hoist.

hydraulic coupling *n:* a fluid connection between a prime mover and the machine it drives; it uses the action of liquid moving against blades to drive the machine. Also called fluid coupling.

hydrodynamic brake *n:* a device mounted on the end of the drawworks shaft of a drilling rig. It serves as an auxiliary to the mechanical brake when pipe is lowered into the well. The braking effect is achieved by means of an impeller turning in a housing filled with water. Sometimes called hydraulic brake or Hydromatic® brake.

Hydromatic® *n:* trade name for a type of hydrodynamic brake.

I

internal-combustion engine *n:* a heat engine in which the pressure necessary to produce motion of the mechanism results from the ignition or burning of a fuel-air mixture within the engine cylinder.

J

jackshaft *n:* a short shaft that is usually set between two machines to provide increased or decreased flexibility and speed.

jaw clutch *n:* a positive-type clutch in which one or more jaws mesh in the opposing clutch sections.

L

live end *n:* the end of a flexible steel brake band that is attached to a brake lever by means of a linkage. Moving the brake lever pulls the live end down, and the whole band tightens around the flange. This slows or stops the drum by friction.

low drum drive *n:* the drawworks drum drive used when hoisting loads are heavy.

M

main brake *n:* two bands fitted with brake pads; the bands fit over the two rims of the drawworks drum. When the driller engages this brake, the pads press down on the rims to stop the drum from hoisting or from letting out drilling line. The main brake also keeps the drum from rotating (and therefore holds the drill stem stationary) when making up or breaking out drill pipe.

makeup cathead *n:* a device that is attached to the shaft of the drawworks and used as a power source for screwing together joints of pipe. It is usually located on the driller's side of the drawworks. Also called spinning cathead. See *cathead*.

manual lubrication *n:* see *drip lubrication*.

mast *n:* a portable derrick that is capable of being raised as a unit, as distinguished from a standard derrick, which cannot be raised to a working position as a unit. For transporting by land, the mast can be divided into two or more sections to avoid excessive length extending from truck beds on the highway. Compare *derrick*.

mechanical cathead *n:* see *automatic cathead.*

mechanical-drive rig *n:* see *mechanical rig.*

mechanical rig *n:* a drilling rig in which the source of power is one or more internal-combustion engines and in which the power is distributed to rig components through mechanical devices (such as chains, sprockets, clutches, and shafts). Also called a power rig. Compare *electric rig.*

micrometer *n:* 1. a caliper for making precise measurements; a spindle is moved by a screw thread so that it touches the object to be measured. 2. an instrument used with a telescope or microscope to measure minute distances.

multistrand drive *n:* a chain-and-sprocket drive in which each shaft has several sprockets, each with its own chain, or strand. The chains are connected to each other, so all the sprockets and chains on one shaft move at the same time and still make up only one drive.

O

off-driller's side *n:* the side of the drawworks opposite the driller. Compare *driller's side.*

offset link *n:* in transmission chain, a combination of roller link and pin link used when a chain has an odd number of pitches.

offset misalignment *n:* a type of chain alignment in which the ends of the shafts, and therefore the sprockets, are not in line with each other. Some sprockets can slide on the shaft, so it is possible for them to be misaligned even if the shafts are not. Offset misalignment alternately stresses the link plates on one side of the chain and then those on the other side. Compare *angular misalignment.*

oil bath *n:* a type of lubrication in high-speed chain-and-sprocket drives in which a portion of the chain passes through an oil bath (sump), which coats all of the chain on each revolution to lubricate it.

oil shear *n:* in a chain-and-sprocket drive, the friction from the chain passing through the oil in the sump.

overrunning clutch *n:* 1. a special clutch that permits a rotating member to turn freely under certain conditions but not under others. 2. a clutch that is used in a starter and transmits cranking effort but overruns freely when the engine tries to drive the starter.

P

peening *n:* in the brake lining, flattening the ends of the bolts to keep the nuts from coming off. Peening does not guarantee that the bolts will stay tight.

pin diameter *n:* in roller chain, about $\frac{5}{16}$ of the pitch. See *pitch.*

pinion *n:* 1. a gear with a small number of teeth designed to mesh with a larger wheel or rack. 2. the smaller of a pair or the smallest of a train of gear wheels.

pin link *n:* a link of roller chain consisting of four parts—two side bars and two pins. The pins are press-fitted into the side bars (pin link plates).

pin link plate *n:* in roller chain, the plate into which the pin link pin ends are immovably fixed. The pins are riveted to the link plate on one side and either riveted or fixed with cotters to the other pin link plate.

pitch *n:* in roller chain, the distance (in inches or millimetres) between the centers of two members next to one another, i.e., the distance between the centers of the bushings or rollers.

plate clutch *n:* a type of friction clutch that has metal plates that may or may not have a friction material on them and an air-operated diaphragm that presses the plates together. Springs push the plates apart when the clutch is disengaged. Compare *drum clutch.*

positive clutch *n:* a clutch in which jaws or claws interlock when pushed together, e.g., the jaw clutch and the spline clutch.

power side *n:* the back of the drawworks, that is, the side nearest the engines that supply power to the drawworks.

pressure lubrication *n:* in a chain-and-sprocket drive, using a pump to force oil continuously through pipes with holes in them, through nozzles, or through holes in the sprockets. Oil sprays out directly onto the chain with pressure. On some rigs, the oil sprays onto the outside of the chain, and on others onto the inside of the chain. Compare *slinger disk lubrication; drip lubrication.*

prime mover *n:* an internal-combustion engine or a turbine that is the source of power for driving a machine or machines.

R

reeve *v:* to pass (as a rope) through a hole or opening in a block or similar device.

roller chain *n:* a type of chain that is used to transmit power by fitting over sprockets attached to shafts, causing rotation of one shaft by the rotation of another. Transmission roller chain consists of offset links, pin links, and roller links.

roller diameter *n:* in roller chain, the outside diameter of the roller, about ⅝ of the pitch.

roller link *n:* one of the links in a roller chain. It consists of two bushings press-fitted into the link plates (side bars) and two rollers that fit over the bushings. The bushings are locked into the link plates to prevent rotation.

roller width *n:* in roller chain, the width of the rollers, which is the distance between the inside faces of the roller link plates.

rotary drilling *n:* a drilling method in which a hole is drilled by a rotating bit to which a downward force is applied. The bit is fastened to and rotated by the drill stem, which also provides a passageway through which the drilling fluid is circulated. Additional joints of drill pipe are added as drilling progresses.

rotary drive countershaft *n:* a rotating shaft, on the opposite side of the drawworks from the driller's console, that gets power from one of the transmissions and sends it to the rotary table. It sits inside its own housing, which may be a part of the main drawworks frame or may be detachable for transportation.

rotary side *n:* see *off-driller's side.*

rotary table *n:* the principal component of a rotary, or rotary machine, used to turn the drill stem and support the drilling assembly. It has a beveled gear arrangement to create the rotational motion and opening into which bushings are fitted to drive and support the drilling assembly.

rotor *n:* 1. a device with vanelike blades attached to a shaft. The device turns or rotates when the vanes are struck by a fluid directed there by a stator. 2. the rotating part of an induction-type alternating current electric motor.

S

sandline *n:* a wireline used on drilling rigs and well-servicing rigs to operate a swab or bailer, to retrieve cores or to run logging devices. It is usually 9/16 of an inch (15 millimetres) in diameter and several thousand feet or metres long.

sand reel *n:* a metal drum on a drilling rig or a workover unit around which the sand line is wound. On a drilling rig, it may be attached to the drawworks catshaft and may be used for coring or other wireline operations. When used on a drilling rig, it is sometimes called a coring reel.

scale *n:* 1. a mineral deposit (e.g., calcium carbonate) that precipitates out of water and adheres to the inside of pipes, heaters, and other equipment. 2. an ordered set of gauge marks together with their defining figures, words, or symbols with relation to which position of the index is observed when reading an instrument.

selective transmission *n:* once the drawworks receives power from either the compound or electric motors, the type of transmission that allows the driller to select how the power is distributed (torque/speed combinations) to various components of the drawworks. The drawworks on both mechanical-drive and electric-drive rigs have a selective transmission.

sheave (pronounced "shiv") *n:* 1. a grooved pulley. 2. support wheel over which tape, wire, or cable rides.

slinger disk lubrication *n:* in a chain-and-sprocket drive, a type of oil bath lubrication in which the chain itself does not pass through the sump, but one or two disks rotating with the sprocket pick up oil from the sump and sling it against a plate, which then feeds it to the top of the lower span. Compare *drip lubrication; pressure lubrication.*

spline clutch *n:* a positive-type clutch that works by means of metal strips that fit into keyways.

sprocket *n:* 1. a wheel with projections on the periphery to engage with the links of a chain. 2. a projection that fits into an opening of a chain.

sprocket ratio *n:* in a chain-and-sprocket drive, the amount of difference in size between the two sprockets.

stator *n:* 1. a device with vanelike blades that serves to direct a flow of fluid (such as drilling mud) onto another set of blades (called the rotor). The stator does not move; rather, it serves merely to guide the flow of fluid at a suitable angle to the rotor blades. 2. the stationary part of an induction-type alternating-current electric motor. Compare *rotor*.

stress riser *n:* a notch or pit on a pipe or joint that raises the stress level and concentrates the breakdown of the metal structure. Also called stress concentrator.

substructure *n:* the foundation on which the derrick or mast and usually the drawworks sit. It contains space for storage and well-control equipment.

sump *n:* a low place in a vessel or tank used to accumulate settlings that are later removed through an opening in the bottom of the vessel.

supply reel *n:* the spool on which drilling line is wrapped as it comes from the supplier.

swivel *n:* a rotary tool that is hung from the rotary hook and traveling block to suspend the drill stem and to permit it to rotate freely. It also provides a connection for the rotary hose and a passageway for the flow of drilling fluid into the drill stem.

T

tong pull line *n:* the catline on the automatic catheads; either a wire rope or a chain.

top drive *n:* a device similar to a power swivel that is used in place of the rotary table to turn the drill stem. It also suspends the drill stem in the hole and includes power tongs. Modern top drives combine the elevator, tongs, swivel, and hook.

torque *n:* the turning force that is applied to a shaft or other rotary mechanism to cause it to rotate or tend to do so. Torque is measured in units of length and force (foot-pounds, newton-metres).

torque converter *n:* a hydraulic device connected between an engine and a mechanical load such as a compound. Torque converters are characterized by an ability to increase output torque as the load causes a reduction in speed. Torque converters are used on mechanical rigs that have compounds.

transmission *n:* the gear or chain arrangement by which power is transmitted from the prime mover to the drawworks, mud pump, or rotary table of a drilling rig.

traveling block *n:* an arrangement of pulleys, or sheaves, through which drilling line is reeved and which moves up and down in the derrick or mast. See *block*.

tugger *n:* a hoist operated by compressed air; a pneumatic hoist. Air hoists are often mounted on the rig floor and are used to lift joints of pipe and other heavy objects.

U   **ultimate strength** *n:* the greatest stress that a substance can stand under normal short-term experiments.

V   **V-belt** *n:* a belt with a trapezoidal cross section, made to run in sheaves, or pulleys, with grooves of corresponding shape.

viscosity *n:* a measure of the resistance of a fluid to flow. Resistance is brought about by the internal friction resulting from the combined effects of cohesion and adhesion. The viscosity of petroleum products is commonly expressed in terms of the time required for a specific volume of the liquid to flow through a capillary tube of a specific size at a given temperature.

W   **water jacket** *n:* in the cooling system, a passageway inside the rim for circulating water.

wear elongation *n:* in a roller chain drive, an amount of stretching that causes the chain to engage improperly with the sprockets.

Review Questions
LESSONS IN ROTARY DRILLING
Unit I, Lesson 6: The Drawworks and the Compound

1. Name five components of the hoisting system.

   (a)

   (b)

   (c)

   (d)

   (e)

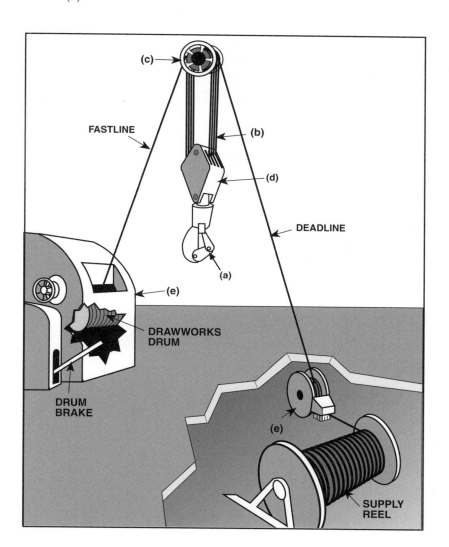

2. Name the major components of the drawworks seen from the front.

(a)

(b)

(c)

(d)

(e)

(f)

(g)

(h)

3. Name the major components of a drawworks seen from the back.

(a)

(b)

(c)

(d)

(e)

(f)

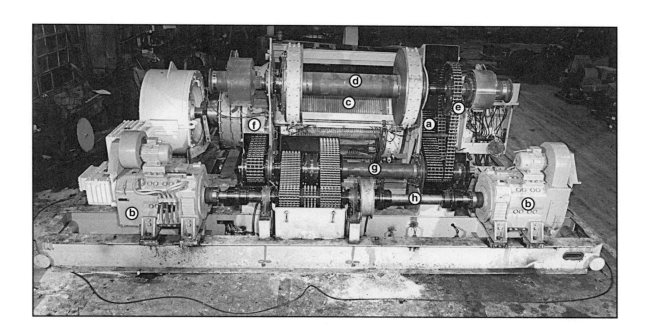

4.   The drum is a big spool with grooves.

   ___ T

   ___ F

5.   A transmission uses drives to transmit rotational energy.

   ___ T

   ___ F

6.   The rotary drive countershaft sends power to the rotary table.

   ___ T

   ___ F

7.   The two types of brakes on the drawworks for the drum are the _____
     brake and the _____ brake.

8.   What are a rig's prime movers?

9.   A mechanical-drive rig uses a _____ to transmit power from the prime movers to
     the drawworks.

10. An electric-drive rig has_____ on the prime movers to generate
     electric power to send to _____ in the drawworks.

11. Name the components of a chain-and-sprocket drive.

   (a)

   (b)

   (c)

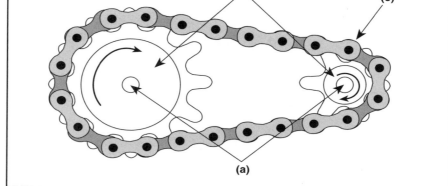

12. The _____ shaft provides power and the _____ shaft receives power.

13. All rigs have a selective transmission.

___ T

___ F

14. Name the drives that make up a chain-and-sprocket selective transmission.

(a), (b), (c)

(d)

(e)

(f)

(g)

(h)

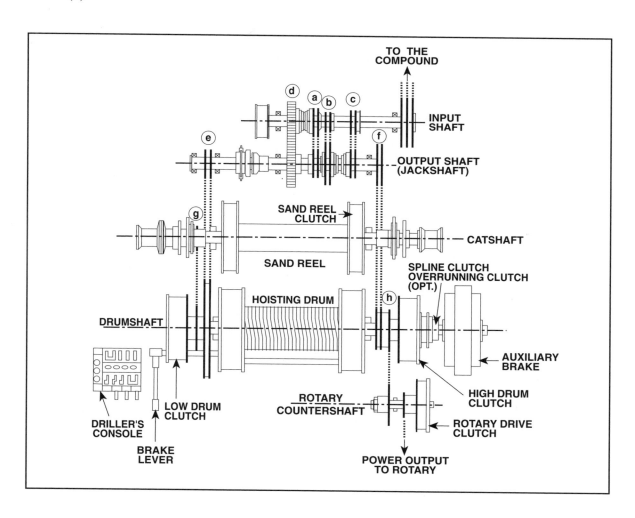

15. Roller chain for transmissions is made up of alternating _____ links and

    _____ links.

16. Name the other two types of links.

17. Roller chain normally has an _____ number of pitches.
    (a) even
    (b) odd

18. When replacing an old chain with a new one, the dimensions and specifications must

    never match.
    ___ T
    ___ F

19. Chain that is too tight or too loose is no problem.
    ___ T
    ___ F

20. Two kinds of misalignment are _____ and _____.

21. _____ oil in the sump of a bath-lubricated drive

    causes overheating.
    (a) Too much
    (b) Too little

22. Change the oil when it is dirty.
    ___ T
    ___ F

23. What does a clutch do?

24. Two types of positive clutches are _____ clutches and _____ clutches.

25. Two types of friction clutches are _____ clutches and _____ clutches.

26. An overrunning clutch differs from other clutches because it works _____ .

27. The most important maintenance job on a friction clutch is to align the driving and the driven shafts.

___ T

___ F

28. Name the three main parts of a mechanical brake.

(a)

(b)

(c)

29. The end of the band attached to the brake lever is the _____ end.

The end anchored to the drawworks frame is the_____ end.

30. The main brake works best when the traveling block is

(a) ascending.

(b) descending.

31. The mechanical brake works best when it is overheated.

___ T

___ F

32. Hard water can cause_____and _____ in the circulation system.

33. Out-of-round brake bands cause the rims and blocks to

(a) wear out prematurely.

(b) fall off.

34. Replace bands, rims, and blocks whenever you have a little spare time.

___ T

___ F

35. Grease and adjust the brake linkage regularly.

___ T

___ F

36.   The auxiliary brake _____ stop the drum by itself.

      (a) can

      (b) cannot

37.   A hydrodynamic brake works through the action of _____ .

38.   An electrodynamic brake works by flipping a switch.

      __ T

      __ F

39.   The most important maintenance job for the auxiliary brake is maintaining the circula-

      tion system.

      __ T

      __ F

40.   The catshaft has two _____ catheads and two _____

      catheads.

41.   The _____ catheads are used to make up and break out

      pipe.

42.   The sand reel is an optional hoist for small tools.

      __ T

      __ F

43.   How do you know when and where to lubricate the drawworks?

Answers to Review Questions
LESSONS IN ROTARY DRILLING
Unit I, Lesson 6
The Drawworks and the Compound

1. (a) Drilling hook
   (b) Drilling line
   (c) Crown block
   (d) Traveling block
   (e) Drawworks

2. (a) Driller's console and brake lever
   (b) Low drum drive
   (c) Main brake
   (d) Drum
   (e) High drum drive
   (f) Catshaft and optional sand reel
   (g) Rotary drive countershaft (optional)
   (h) Auxiliary brake

3. (a) High drum drive
   (b) Electric motors
   (c) Drum
   (d) Optional sand reel
   (e) Catshaft drive
   (f) Low drum drive
   (g) Output shaft
   (h) Input shaft

4. T

5. T

6. T

7. mechanical; auxiliary

8. The basic power source; diesel-powered internal-combustion engines

9. compound

10. generators; motors

11. (a) Shafts
    (b) Sprockets
    (c) Roller chain

12. driving; driven

13. T

14. (a), (b), (c) Chain-and-sprocket drives for three speeds
    (d) Gears for reverse speed
    (e) Low drum chain-and-sprocket drive
    (f) High drum chain-and-sprocket drive
    (g) Chain-and-sprocket drive to the catshaft
    (h) Chain-and-sprocket drive to the rotary drive countershaft

15. roller; pin

16. connector; offset

17. even

18. F

19. F

20. angular; offset

21. Too much

22. T

23. It connects and disconnects two rotating shafts.

24. jaw; spline

25. drum; plate (disk)

26. automatically

27. T

28. rims (flanges); bands; blocks

29. live; dead

30. descending

31. F

32. scale; corrosion

33. wear out prematurely

34. F

35. T

36. cannot

37. water

38. T

39. T

40. friction; automatic

41. automatic

42. T

43. Drawworks manufacturers provide charts of recommended maintenance schedules, showing where to lubricate and how often, and what type of lubricants are best.